Astronomers' Universe

For further volumes:
http://www.springer.com/6960

Other titles in this Series

Michael Perryman

The Making of History's Greatest Star Map

 Springer

Cover images : The Hipparcos mirror integration (ESA, 1987), and the satellite launch from Kourou in August 1989 (courtesy ESA/CNES/Arianespace). The background is a synthetic image of the Hyades star cluster, constructed from the Hipparcos catalogue by Jos de Bruijne.

ISBN 978-3-642-11601-8   e-ISBN 978-3-642-11602-5
DOI 10.1007/978-3-642-11602-5
Springer Heidelberg Dordrecht London New York

Library of Congress Control Number: 2010922053

Cover design: Mary Ann Smith, New York

Printed on acid-free paper

Springer is part of Springer Science+Business Media (www.springer.com)

For my mother, who pointed me on the journey of discovery.
And for all who contributed to the making of history's greatest star map.

Preface

O UR UNIVERSE IS LARGE AND COMPLEX. To contribute to its understanding a satellite was launched by the European Space Agency in 1989 with a unique if prosaic brief: to chart the positions and motions of the stars.

History's greatest star map was the result. Since its completion in 1997, astronomers worldwide have devoured and pondered its content. They have used it to explore the formation and final demise of stars, and to peer at the nature of nuclear reactions deep within them. They have studied minute changes in star positions to trace how our Galaxy came into existence billions of years ago, and used their distances to date the age of the very Universe itself. They have consulted this detailed map of the heavens to investigate changes in Earth's climate over the last hundred million years. They are sifting its content to identify future impacts of 'death stars' heading towards our solar system, and to have a better fix on the likelihood of other habitable planets in the neighbourhood of our Sun. They have used star positions of exquisite accuracy to test Einstein's theory of general relativity with greater precision than ever before. That we can even hope to divine such knowledge from the careful measurement of star positions from space deserves some explanation.

The United Nations declared 2009 to be the International Year of Astronomy, focused on helping citizens of the world discover their place in the Universe through the majesty of the night-time sky. Twenty years after launch, and a decade after the catalogue release, it's a good time to relate the story of an unusual space mission: the foundations laid by those who went before, its planning and its near failure, and the indelible impression it has left on our understanding of the Universe.

THE SATELLITE AT THE CENTRE OF THIS STORY, HIPPARCOS, and its great star map, are well known to the professionals. The idea of crafting a more popular account, something that could explain a large and complex space programme to those more usually excluded from these things, had been in my mind while I worked on the project. It assumed a greater urgency as the grand scale of the undertaking unfolding around us became more apparent, and as I became more aware of the subject's remarkable history.

Like so many good intentions it took a back seat, first to the satellite's successor which I worked on for a further decade. After that, I turned back to a task tied to Hipparcos which had also long been in my thoughts—to examine the scientific literature resulting from the use of the catalogue. For in the ten years since the results had been published, scientists around the world had been busy decoding their meaning, and I was curious to see what investigations had been made. I had the opportunity to discover the splendid things that astronomers had been able to do with the star positions that our satellite had measured. I read many scientific papers, and wrote a detailed text which looked across the entirety of its investigations, steering a guide for those continuing to research the field.

Many of the new results surprised me, a large number enchanted me, and most brought to a much clearer focus exactly why we had worked on this project for so long. From this modest-sized telescope in space flowed results which have given science many new advances. Some of the investigations are quite esoteric. But there were many unexpected discoveries, and plentiful new insights, which I knew I would like to explain. There are many curious things being unravelled about our Universe that the non-specialist might like to know.

An army of collaborators had been needed to make this most advanced map of the stars: the satellite project team and the flight operations team, the launch authorities, and the advisory committees and steering committees and funding committees. The scale of the effort was impressive: the industrial organisations across Europe numbered around thirty, and the total number of people implicated probably around two thousand. Central to the whole venture were the scientific teams—in total around two hundred individuals, of different nationalities, with their different heritage and their subtly disparate ways of doing business. The project, like all space ventures, was difficult and it was relentless, with plenty of highs and its fair share of lows. But, from all sides, it represented a remarkable collaboration. It took me to many countries, and I worked with many people. All were committed to the same end result. All were caught up in the project's energy, carried along towards a clear end goal—a charting of the heavens. It was a wonderful collaboration, and I wanted to emphasise that human aspect too.

As time progressed I became more conscious of the work of those who had gone before, and who had laid the foundations of the field. At first, new to big science, I had the impression that we were working at *the* forefront of astronomy, grateful that we had not had to struggle with the inadequate tools of the past. Progressively it became obvious, of course, that though we were at the cutting edge, it was but a temporary one, as it was for our predecessors, and as it would be for those that will follow. Our task was challenging, and very exciting. And I realised, eventually, that theirs had been too. I became far more aware of the gratitude that we owe to those that had pushed back the frontiers before us. Conscious also that we should take time, and care, to comprehend the height of the hurdles over which they had climbed, and the leaps of imagination they had needed to progress throughout the centuries that have passed before us.

In researching the history of the field, I got hold of a copy of *The Great Star Map*, which inspired the present title. Written in 1912 by Professor H.H. Turner from Oxford, the book is a first-hand account of another huge celestial survey, the Astrographic Catalogue. It is an engaging description of a substantial and quite remarkable scientific collaboration undertaken a century ago. The language is quaint. Descriptions of the meetings, and how the protagonists set about their tasks, are truly those of another age. The Astrographic Catalogue too was once a grand undertaking at the leading edge of science, but the details are already all-but-lost in the mists of time. *The Great Star Map* is a record of an international enterprise lasting almost half a century, undertaken with tools almost too primitive for us to imagine. It is the story of vast quantities of star images measured by eye, and of mountains of numbers meticulously recorded and laboriously computed by hand.

The human side of the Hipparcos satellite project will be lost quickly too. The catalogue will be remembered for sure, and along with it the conceptual leap involved in making the observations for the first time from space. But the experiment, and how it was assembled and launched and operated, the monumental computations, and the way of doing scientific business in space at the turn of the second millennium, all too will soon be swept aside, relegated to little more than a curiosity of the past. So this is also a record of our own 'great star map'.

It is fashionable in popularising science to introduce the characters and weave much of the story around them. I have nothing new to add about the temperaments and characters of the great achievers in this field from the past. I have looked over its history, and extracted the pieces of the developments which are most relevant for this story. Of the men and women who played their part in Hipparcos, happily most are still alive, and I have neither the wish nor the skills to dissect their motivations or temperaments.

I wanted instead to record the scientific, technical and organisational challenges, and how we set about conquering them, and to describe a few of the astronomical results and something of what they mean.

For seventeen years I led the scientific side of a complex project. If I describe parts from a personal perspective, it was because I was there as a witness, not because my experiences are any more relevant than those of others. I sat in the middle, but each of us has our own story. What is important is that we picked up the baton of science together, ran with it as hard as we could, and placed it firmly in the hands of those next on an endless journey of scientific discovery.

THE UNIVERSE IS BEYOND COMPREHENSION, but our place within it deserves to be explained as best we can. And while astronomers generally pretend limited knowledge of direct practical value, we can see with sometimes startling acuity that humanity is huddled together on a tiny and desolate rock hurtling through space, divided from this unsettling reality by artificial boundaries of nationality and beliefs. If we could articulate that insight more clearly, it might be a tiny but definable contribution towards a common spirit of awed existence.

Michael Perryman
December 2009

Contents

Prologue: Hipparcos Launch

Space isn't remote at all. It's only an hour's drive away if your car could go straight upwards.

Fred Hoyle (1915–2001)

I N A CLEARING OF THE TROPICAL coastal forests of French Guiana in South America sits the launch site from which Ariane flight 33 powered into space on the sultry night of 8 August 1989. Atop the sixty meter high rocket was one of the most advanced, delicate, and expensive instruments in physics and astronomy ever conceived. Designed, built and operated under the financial and technical might of the European Space Agency, the Hipparcos satellite was the culmination of more than twenty years of brainstorming by large scientific and technical teams, of political negotiations and financial lobbying, and of meticulous engineering design, construction and testing. Now it sat enclosed, high in its metallic eyrie, alone for the first time, and ready to ride into its place in history.

At the time Europe's largest and most powerful launch vehicle, the Ariane 4 rocket was produced and operated by Arianespace, a French-led company whose roots lay in plans for European autonomy in space transportation drawn up in the 1970s. Its launch site dominates French Guiana, an overseas department of France which, being part of the European Union's territory, is even subtly depicted on all euro bank notes. The launch centre at Kourou, some fifty kilometers from the capital Cayenne, employs two thousand people, and accounts for a large fraction of the department's economy.

M. Perryman, *The Making of History's Greatest Star Map*, Astronomers' Universe
DOI 10.1007/978-3-642-11602-5_1, © Springer-Verlag Berlin Heidelberg 2010

Jutting out into the Atlantic Ocean, way to the east of New York, French Guiana lies on the north-east edge of the vast Amazon basin. Here, even small and insignificant tributaries dwarf all but the largest European rivers. Impenetrable forests sustain an ecological kingdom which screams into life every evening as the tropical darkness slices through the uncomfortable humidity of the day.

Its existence as Europe's space port hangs on its auspicious location. Being on French soil simplifies the politics if not the logistics, and being close to the equator is a big advantage for satellites being lifted into geostationary orbit. The area is relatively immune from hurricanes and earthquakes, and its location on the north-east coast of South America ensures that launches can be directed out over the sea with relative safety. The nearby coastal area, including the infamous Devil's Island, a penal colony until 1953 and the setting of Henri Charrière's celebrated narrative novel *Papillon*, is evacuated during launches as a further precautionary measure.

THE HIPPARCOS SATELLITE, around which this story unfolds, had been transported from Europe several weeks previously, under the overall authority of project manager Hamid Hassan. Unique, delicate, and costing considerably more than its weight in gold to develop, it received, accordingly, all the attention and care of a premature baby.

Its transport by road from Torino in Italy to Toulouse, and then on to French Guiana, had demanded a special protective casing constructed by the shipbuilding firm of Sener in Bilbao. Inside that, the satellite crossed the world within a large pressurised rubber balloon made by inflatable boat builders Zodiac, carefully designed to keep contamination at bay. Some way on its snail-like journey through France, the gas valve had stuck open and the giant balloon had duly ruptured, but the convoy included engineers with a suitably scaled-up puncture repair kit, and no lasting damage was done.

Transport from Toulouse

From Toulouse, an Air France cargo flight, which had demanded twelve hours of the most delicate supervision to load, flew out of Blagnac airport carrying its precious cargo to Cayenne. Uncrated and re-tested, it would finally be tucked away inside the launcher's protective nose cone, the whole rocket then to be inched out from the assembly bay to the launch pad.

This pioneering satellite would then be poised to embark on its final journey to map the stars from space.

The previous weeks in Kourou had not all been plain sailing. Sequential but overlapping launch campaigns meant that four different groups were operating at the same time, overstretching the limited local facilities. Several of the Hipparcos team were quartered for weeks on a boat annexed by a local hotel, with cramped cabins and a noisy power generator for company. One who put his foot down at these unsuitable lodgings was the man in charge of final preparation and verification activities, quick-fire Italian Oscar Pace. Arriving for his third time in Kourou for the final stage of the launch preparations, Pace was dropped from his taxi at the nearest point on the road a hundred meters distant. The tropical heavens opened half way on his dash to the boat, and he stepped onboard drenched, bedraggled and not a little unhappy. With the unremitting humidity and primitive facilities, he and his clothes were still soaked through the following morning. "If you want your verification engineer to verify anything," he told Hassan uncompromisingly, "find me a real hotel!"

An intensive schedule of final preparations, tests, and launch rehearsals dominated the team's attention in Kourou for several weeks. All the necessary steps had been planned and timed with the greatest care. But the intensive work schedule was punctuated by moments of relaxation, of course, and visitors are always anxious to see a little of the local colour. Colourful restaurants offered a choice of the exotic creole cuisine, with shark steak and caiman for the regular tourists, agouti and macaque for the more adventurous, and the innocuous-looking cayenne pepper poised to inflict its explosive kick on the more naive,

Oscar Pace (2007)

or the more heroic, and there were always a few of each prepared to try.

Tour companies offered trips into the jungle interior for the curious. Hamid Hassan and Oscar Pace patched up their earlier clash over accommodation by taking an escorted dug-out canoe together up the Kourou River and into the forest, winding in and out through a maze of side creeks which, they said, had looked exactly the same for mile upon mile, and in which they soon became completely disoriented. Hassan, an Englishman of Pakistani descent who occasionally dropped dark hints of a noble background and of periods still spent in Kashmir hunting on horseback, admitted to feeling distinctly nervous should their guide, unequipped with map or radio communication, have had an accident. Pace, always quick to see the droll side of any situation, countered that he personally had no such concerns, and indeed had a distinct advantage: he was, he said, a believer—and was already praying with the utmost fervour.

Dieter Bock, the project's chief mechanical engineer, took off for a week's trek into the deeper interior and was not so lucky. Returning with stories of ferocious mosquitos that penetrated even through the protective netting, he contracted a serious case of malaria, and was flown back to Europe for intensive treatment.

A last-minute component quality alert from industries in Europe added to the routine stresses of the final preparations. The news had scrambled an urgent sequence of unplanned tests, tense discussions, and risk analyses that had dragged on for days. Hassan looked over the final reports which summarised the known facts, then quickly took the decision to proceed. His eyes squinted through the smoke rising from his own cigarette. "There is no problem, we launch," he said dismissively. Such were the decisions he was paid to make.

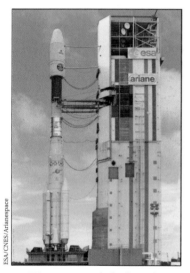

ESA/CNES/Arianespace

Hipparcos ready for launch

Sharing the journey into space, and the not inconsiderable €100 million launch fee[1], was the French–German television broadcasting satellite TV–Sat 2. With the component alert out of the way, another, more literal, thunderbolt hit as the two satellites sat side-by-side in the integration building. A lightning strike, common enough in the tropical thunderstorms, leapt from a conduit to an injudiciously placed ladder, and fried the low-voltage circuitry of the co-passenger's test equipment. It called for an unscheduled launch delay of a further fortnight, although Hipparcos had escaped intact.

The hiatus meant hurried telexes to Europe—this all took place in the pre-email era—and an anxious bustle of flight and hotel rescheduling for those still due to fly out to see the launch. For those waiting it out in French Guiana, it meant a few more enchanted mornings seeing the sun rise over the Atlantic, and a few more afternoons watching rain squalls chasing across the forest canopies.

A previous near-miss had happened a year earlier, when the satellite had been sitting in a test chamber in the Alenia premises outside Torino in Italy. Test engineers had heard an explosion, and stepped outside to find two smoldering craters right by the wall of the building, fifteen centimeters deep and a meter apart. A meteorite had crashed to the ground, missing the building by a whisker. The story raced around Europe, garbled and amplified. Even the most rational had pondered its significance.

TWO TENSE FINAL STEPS were needed to ready the satellite for launch. The first was to fill its storage tanks with the highly combustible hydrazine propellant which would position the satellite precisely once in space. Hydrazine is a colourless liquid that looks like water but smells of ammonia. It is, however, dangerously toxic and highly unstable. Its components of nitrogen and hydrogen split apart with the gentle prod of catalysts to produce a large volume of hot gas from a small volume of hydrazine liquid, making it a particularly efficient thruster propellant. Its power unmatched, it was used for the first rocket-powered fighter plane, the Messerschmitt 163B, and for the touchdown rockets of the Viking programme's Mars landers in the 1970s.

Filling the satellite's hydrazine storage tanks was a slow and nerve-wracking event, undertaken by the suitably cautious experts from industry. So dangerous was the fuel that the attendant staff was reduced to a strict minimum, with those that did have to be there wearing full protective breather suits in deference to its toxicity. They trickled the innocuous-looking liquid slowly into the tanks over a period of several days, cleanliness at a premium to preclude contamination of catalyst and nozzles. To minimise any possible risk of sparking, they tip-toed around cautiously, with grounded straps and special air-cushioned shoes, and even wrist watches with their tiny batteries were prohibited. The local hospital was placed on standby as a further contingency.

As if this dangerous fuel was not quite enough, the satellite carried one further jumbo firecracker—half a tonne of solid rocket propellant packed into a boost motor and bolted to the satellite, providing a mighty push which would lift it into its final circular orbit once in space. The control centre activated its red-flagged safe-and-arm device as part of the final countdown sequence. The solid boost motor was thereby primed for later ignition by a radio command that would be issued from the ground to the satellite when high in orbit.

In one bundled high-tech package, delicate optics, a battery of sensitive instruments, and a highly explosive arsenal would shortly ride into space together. And so it was that, with all preparations at last completed, Hipparcos was finally ready for launch.

THE LIFT-OFF MASS of this particular Ariane rocket, tailored to its specific and demanding dual launch with three liquid stages and two additional solid rocket strap-on boosters, was a little over four hundred tons. Its initial thrust, technically rated at more than five million Newtons, was twenty times that of the 1997 world land speed record holder *Thrust SSC*. I had gazed and reflected on this sixty-meter tall towering giant in the assembly bay, some weeks before. It seemed almost absurd to me, and certainly defying intu-

ition, that the intricate delicacy of the Hipparcos satellite should be consigned to the massive and potentially destructive acceleration and vibra-
tional forces that would embrace it during its launch. Clever engineering
design and analysis, and the brute-force shaker tests carried out in Europe
to prove the mechanical models, would hopefully ensure its survival.

Engineers and managers from the European Space Agency, and from
the various industrial teams that had been involved in the satellite's creation,
were here to see the spectacular lift-off of this space-age star-measuring machine. I was there in my capacity as ESA's chief scientist for the project, a
position I'd held for the eight years of the satellite's gestation. A dozen European scientists were there to watch its departure; they represented many
more who had devoted years of their professional lives to this unusual venture. Dignitaries from the senior advisory and funding bodies of the European space programme completed the spectator line-up.

Twenty years ago, passenger flights in and out of French Guiana were
restricted in number, and the occasional launch campaigns placed a difficult strain on international flights as well as on scarce hotel rooms. Most of
the invited spectators, some seventy in number and representing the ESA,
Hipparcos and TV–Sat 2 communities, arrived on a specially chartered Con-
corde flight from the national departures terminal at Charles de Gaulle airport in Paris. They would be flying out again, as would others arriving on the
scheduled carriers, two days after the planned launch date, whether or not
the launch took place.

THE LAUNCH SEQUENCE WAS BROADCAST to the invited spectators sitting in
tiers behind a massive glass panel at the back of the Jupiter control room, itself located for safety reasons several kilometers from the launch pad. I took
my place between the original proposer of this mission, Frenchman Pierre
Lacroute, then aged 83, and the chair of ESA's Science Programme Committee, Swedish astronomer Kerstin Fredga. On the screens at the front, beyond
the banks of operations consoles of the launch team, was a relayed view of
the majestic Ariane rocket. We could watch the screen and hear the commentary, or slip outside to see it, live, across the forest clearing. The intense
tropical darkness had fallen, and the rocket, lit up by floodlight, was the centre of all attention. From various filler valves it was gently issuing wisps of
evaporating cryogenic fuel, signaling its readiness to depart.

Even from a distance, the rocket appeared immensely tall, but almost
pencil thin. Gantry arms forty meters above the ground supported it from
the launch tower. A couple of dozen power, fuel, and communications lines
hung loosely between the two, life-supporting umbilicals soon to be cut
loose in the seconds before launch.

The rocket itself was a feast of complexity. The two huge twenty-meter tall strap-on boosters were almost lost against the vastness of the main rocket shaft. Around the base, the great exhaust nozzles pivoted on giant gimbal bearings to provide directional control. The cylindrical shaft tapered half way up, before bulging out again at the top to host the nose cone containing the two satellites. And all the way up were the proud logos proclaiming the expert teams involved—Arianespace and CNES as builders and operators of the rocket, with the colourful European flags of the participating nations emblazoned above; higher up, the blue ESA thumbprint-like sign and the launch number 'V33' stamped over an aluminium-like casing. At the very top, on the outside of the nose cone, were the project emblems of the Hipparcos and TV–Sat 2 satellites, now tucked up inside.

The screen in front of us also showed the array of status indicators charting the progress of the countdown, and a close-up of the Arianespace operations director Jacques Bouchet, presiding over his command console. Over the loudspeakers, his voice gave overall coordination to the highly trained launch team. Weeks of specific preparation for this launch, years of steady development of the Ariane family itself, a decade of design and assembly of a telescope to map the stars, all were at last coming together in this singularity of time and space: our own 'big bang'.

Jupiter control room, Kourou

ESA/CNES/Arianespace

Tension mounted as the countdown progressed through a series of synchronised steps. At two pivotal 'go – no go' decision points, at eighteen and then at six minutes before launch, the array of green lamps illuminating the status board continued to reassure us that all launch systems were ready. For the scientists who had played a central role in the satellite design and who had, over several years, made careful preparations for the analysis of the data once in orbit, their brainchild had been irretrievably placed in the hands of the Arianespace launch authorities. For now, all they could do was watch, wait, and hope with fervour that the launch team had done its preparatory work as meticulously as they had done theirs.

"Here's a wonderful image of the Ariane 4 standing sixty meters tall on its launch pad, lit up in the night", proclaimed the launch commentator, who would be keeping us up-to-date with the series of events unfolding before our eyes. "Next to it, a clean, clear green status panel with all systems go. As we pass the eighteen minute mark, we're counting down to the launch of flight 33. It's a hot humid night here in Kourou, but lovely conditions for a launch." And after a pause, "Here in the Jupiter control centre we've just had another clear 'go' on the weather. And here's another fine shot of the launcher as we approach the six minute mark."

The camera switched to a full screen image of the operations director concentrating fiercely on his task of conducting the campaign. Then the first of his countdown sequences was suddenly upon us: «Attention pour le début de la séquence synchronisée: trois, deux, un—moins six minutes.» Then, as our eyes darted between the activities playing out in the control room before us, and as our pulses raced and adrenaline surged: «Trois, deux, un—moins une minute.»

Arianespace/Jacques Bouchet

Jacques Bouchet

The commentator continued: "The final conditioning of the launch vehicle systems is taking place, and at minus five seconds we'll see the cryogenic arms open." Expectation intensified. And at fifteen seconds from launch the voice of the operations director once again: «Attention pour le décompte final... 10, 9, 8, 7, 6, 5, 4, 3, 2, 1...»

But silence reigned. The image of the launch vehicle remained unchanged, stubbornly reposing on its launch pad. The status screens suddenly flashed up unwanted banks of red indicator lights. It was difficult to imagine what had gone wrong, but the operations director soon calmly announced the obvious: «Absence d'ouverture des bras», swiftly followed by the commentator's: "Well, it looks like the clamping arms have not opened. For some reason the computer has decided that it would not authorise the opening of the arms; therefore no take-off occurred."

A SATELLITE'S LAUNCH WINDOW can be surprisingly restricted, as short as a couple of hours on a given day. This, in turn, may be a window lasting only a few days of the year, depending on the orbit and location that the satellite has been designed to occupy. It may sound more like astrology than astrophysics, but the relevant celestial bodies must be properly aligned for gravity to do its work and help position the satellites as planned. TV–Sat 2 also had to be launched into a special position, such that Sun eclipses, which would cause a temporary drop in its solar array power, had to avoid the periods of peak energy demands—the times of the evening news over Europe. There was less than an hour to go before this dual launch attempt would have to be aborted and postponed to another day.

After a few minutes of waiting anxiously, Charles Bigot, Director General of Arianespace, appeared clutching a microphone. He confidently explained that his team had already recognised the problem, and that they would be moving swiftly to try to re-start the launch attempt. We would learn later that the correct swiveling of the vast exhaust nozzles of the three colossal Viking engines, which together determine the rocket's trajectory, had not been confirmed in the status reports, and that the countdown sequence had therefore gone into an automatic hold. My own disappointment, was compounded by thoughts of the frustration and logistical chaos that this would create: some invitees would have to return to Europe the next night for other immovable commitments, others would try to secure at short notice the few remaining hotel rooms in this remote part of French Guiana, and there would be an ensuing scramble for alternative flights back to Europe at a later date.

Charles Bigot

In the remaining thirty minutes of the possible launch window that night, the Arianespace engineers worked quickly to overcome the problem and, amidst palpable relief, the sequence resumed with its synchronised ten-second countdown at minus six minutes.

Witnessing a launch is exciting for any spectator fortunate enough to see it live. But it is doubly nerve-wracking for those whose careers have been invested in the cargo, and whose futures are tied to the launcher's success. Despite the most careful preparations, things do go wrong in such hugely complex systems. Entire projects, years of preparation, and scientific careers ride on the uncertainty of the following few seconds. Disasters had happened before. Originally due for launch in 1988, the Hipparcos satellite had actually sat for a year in storage after the third stage of flight 18 in 1986 had failed to ignite. Launches were delayed while the reasons for the failure were investigated. Some years later, Europe's multi-spacecraft Cluster magnetospheric mission was destroyed on the maiden flight of Ariane 5, which disintegrated thirty seven seconds after lift-off in 1996. In manned programmes, even lives cannot be guaranteed immunity from technical or human failure. Scientific and technical disasters can be accompanied and overshadowed by human tragedy. Launches still represent something of a lottery, and the stakes are always high.

Eventually, the clock ticking towards its deadline, Jacques Bouchet was back on screen with an even more tense repeat countdown. Hunched over the microphone, he wriggled nervously, all eyes upon him. His voice echoed around the hushed control room: «Trois, deux, un—mise à feu!»

THIS TIME, THE BASE OF THE LAUNCHER ignited in a spectacular multiple flash, before subsiding, then rapidly but more uniformly erupting into its full glory. Its engines lit up the night sky, then smoldered and seemingly hesitated for a second or two before this imposing juggernaut rose, surprisingly slowly and sedately, from its launch pad. Its herculean struggle against Earth's gravity had begun.

ESA/CNES/Arianespace

Hipparcos launch, 8 August 1989

The roar of sound waves from the rocket motors crashed across the forest clearing only to reach the Jupiter control room several seconds later. We were guided through the ensuing action by the voice of our commentator. "We have a wonderful shot of Ariane 4 lifting off from its launch pad, and roaring into the night here above Kourou, the six Viking engines performing flawlessly," he reassured us.

I don't know if the next words were scripted, but they were uttered with awed emotion: "It is a beautiful sight!" We sat enthralled at the back of the control room, looking first at images of the launch, later replaced by the infrared heat-sensing tracker capturing the progress of this immensely powerful rocket as it moved gracefully, relentlessly, up and across the night sky, and eventually out of sight.

A ROCKET ASCENDING INTO SPACE is fighting the force of Earth's gravity trying to pull it back down. Gravity is one of the fundamental forces of Nature, and it occupies a central role in the astronomical investigations made by Hipparcos. It is the irrepressible force that holds our feet to the ground, and pulls swirling gas clouds into the dense concentrations from which stars are born. It is the celestial master that dictates the paths of the planets around our Sun, and the endless journeys of all other stars through space.

For a rocket overcoming the Earth's pull, mention is sometimes made of the 'escape velocity'. This is the speed that any object would need to escape Earth's gravity if projected from the surface in a single push, and with no further impulsive force provided along the way. Escape velocity is a little more than eleven kilometers per second at the Earth's surface. In practice a rocket never reaches such a high speed, for as long as it is propelled continuously upwards, it can escape at any velocity. The huge size of satellite-carrying rockets is determined by the weight of fuel that they must carry in order to continue to lift the remaining weight ever upward. Calculating the fuel needed, the mass of the tanks, and the optimum weights of additional boosters is a tricky business.

In the early seconds after launch, the two bright exhaust plumes of the solid boosters dominated the screen image in front of us. Having served their brief but crucial task of providing extra lift, they were jettisoned fifty seconds into the flight, their colossal work taken over by the liquid engines. Three and half minutes into the flight marked the burnout of the first liquid stage, and nearly six minutes into the rocket's flight, now moving at five kilometers a second, came the extinction and separation of the burnt-out second stage. Every so often, we heard the wonderfully reassuring words from the operations director: «Ascension normale, pilotage normal.»

THE LAUNCHER'S TASK was not yet completed, however, and there were two crucial steps which still lay ahead. Both satellites aboard, TV–Sat 2 and Hipparcos, remained sheltered and pinioned within the nose cone, Spelda. Inside, its precious cargo was still being escorted higher by the rocket's third stage. It was a long wait until, twenty two minutes into the flight, our commentator continued: "There is still quite some tension on the faces of the people responsible for Hipparcos at the European Space Agency, Dr Hamid Hassan and David Dale." Barely a minute later, the operations director announced with just a hint of excitement «Séparation Spelda!» This important moment was greeted with a small nod and the briefest of 'thumbs up' from project manager Hamid Hassan.

Our commentator continued: "We have separation of the upper part of Spelda. Hipparcos is now almost in the open. Before the satellite can be released, there will be a re-orientation and spin-up of the launch vehicle." The spin-up meant that Hipparcos would have an additional dynamical stability, just like a spinning top, allowing it to remain pointing in the right direction during its release. Bouchet's voice came in over the top: «Mise en rotation du composite», and the commentator continued, mirroring our own thoughts: "This is a wonderful moment for Europe and the European Space Agency."

Hamid Hassan (1989)

Nearly twenty four minutes after lift-off, at a height of several hundred kilometers above the Earth, moving at nearly ten kilometers a second, and already racing far out across the Atlantic, Bouchet proclaimed the magical words «Séparation Hipparcos!» His words were spoken with a calmly triumphant tone. Tension evaporated.

David Dale, overall head of the scientific projects department in ESA's research and technology centre in The Netherlands, had been rocking gently in his chair in the control room, concentrating intently on the complex launch steps unfolding. He turned and gave a benevolent smile to Hassan.

Both in their early-fifties, with long and successful careers in space already under their belts, this launch was still an enormously pivotal, make-or-break event. In that split-second, Dale communicated relief, pride, and congratulations for the culmination of years of dedicated team work. The two swung around, and gave a brief wave to acknowledge the spontaneous applause from the audience behind the glass screens in the control room.

Our commentator signed off: "That's it! Flight 33 of the Ariane launch programme has just successfully launched Hipparcos. There is jubilation here in the Jupiter control room, and great relief on the part of the Hipparcos team." Almost prophetically, bringing us back to the realities still ahead, he added "Just a reminder that Hipparcos has many things left to do: thirty seven hours into the launch a boost motor will place it in its final orbit, and the solar arrays will only be fully deployed in about thirteen days from now."

THE SUCCESSFUL LAUNCH WAS RELAYED around the world. It was beamed live to the European Space Agency's two-thousand strong research and technology centre tucked into the dunes on the North Sea coast near Noordwijk in The Netherlands, and to the organisation's head office in the rue Mario Nikis near the Eiffel Tower in Paris where its Director General Reimar Lüst was following events. Most crucially, it was relayed to ESA's control establishment in Germany, the European Space Operations Centre, ESOC, on the outskirts of Darmstadt. Here, the operations team, the project's counterpart to the launch contingent in French Guiana, would pick up the signals from the satellite as it hurtled towards Europe, using its network of ground stations spaced around the Earth. Like a high-tech relay race, they would quickly take control of its operations, and of its future.

Meanwhile astronomers around the world, colleagues and families, journalists and politicians, were rejoicing. Emotional tears from some signified the awe felt by all in witnessing the almost incomprehensible power required to free an object from the gravitational embrace of the Earth, and to place it in orbit above the atmosphere. The launch was a testimony to generations of scientific and engineering advances laid down before us, leading inexorably to this latest chapter of astronomical history now unfolding. One critical step for this unusual experiment had been successfully taken.

SOME OF THE PROJECT TEAM had watched from the roof of a control bunker just two kilometers, and directly down range, from the launch pad. The flaring rocket, the size of six double-decker buses nose-to-tail, had thundered above their heads as it headed for Europe. "Supernatural", accompanied by an incredulous shake of the head, was one of a number of similar reverential reactions from those who had seen it depart from these closest of quarters.

Arianespace laid on a celebration party for its delighted customers, held in the gardens of the Hotel les Roches in the South American jungle. Some one hundred and fifty scientists and engineers, and representatives of funding authorities and of space science's highest advisory councils, strolled the gardens, sipped rum punches, intercepted the passing buffet trays, and congratulated each other on a job well done. Hamid Hassan, the focused and charismatic project manager, circulated almost deity-like through the party, the personal embodiment of the surreal event that we had just witnessed. Whisky in one hand and cigarette in the other, he had a personal word of thanks for all of the many players in his satellite team.

Roger–Maurice Bonnet was there as overall direc-tor of the agency's science programme since 1983, the top position in European space science which he was to occupy for another twelve years. But even the busiest agendas make room for a launch. It is the make-or-break moment for each space mission, the instant when the hard work of the past years comes together, and when all the delicate bits and pieces making up an instrument leave the Earth forever. It was, too, an important show-case event for politicians and funding authorities, with influential figures such as the celebrated French Minister of Research, Hubert Curien, also there to judge the spectacle.

Roger–Maurice Bonnet (2000)

Bonnet's job came with considerable authority and a heavy responsibility; he had to cajole delegations to support all the disparate parts of the science programme, and shepherd consensus amongst the various countries. The burden of success fell heavily on his shoulders too. A spokesman for each of ESA's space science missions, Bonnet was a passionate supporter of this cerebral addition to Europe's scientific portfolio. The launch was, for him also, a moment of great pride and massive relief. An articulate and inspiring speaker, he addressed the open-air assembly, and described Hipparcos for us as a "project unique in the history of space missions." The party continued long into the balmy tropical night.

BUT THE ENSUING DAYS, weeks and months would turn out to be far from a smooth journey. While a launch is the most dramatic and precarious single event for most space missions, enacted in the full glare of the media, and of a wider scientific and public scrutiny, many other pitfalls invariably lie along the lengthy and convoluted path of a space mission's life. And a massive setback lay in wait for us.

Chapter 1

Our Place in the Cosmos

I don't pretend to understand the Universe—it's a great deal bigger than I am.

Thomas Carlyle (1795–1881)

I N 1980, THIRTEEN YEARS AFTER THE IDEA for a space experiment to measure the positions of the stars had first been mooted, the international advisory committees of the European Space Agency had assembled and, after lengthy debate, duly accepted Hipparcos as the organisation's next major science programme. In one ambitious leap, this novel experiment would carry the ancient science of measuring star positions well and truly to the forefront of the space age.

Its selection caused quite a stir. Some from the professional community who had watched from the sidelines as the discussions unfolded were fulsome in their praise for the bold choice that European science policy leaders had made. Others were not at all convinced that further mapping of the stars would bring great rewards; for cosmology and studies of far-flung galaxies was where much of the action in astronomy had recently shifted.

To get this far along its road to reality, lengthy studies had been needed to assess the satellite's feasibility and to ensure its accuracy. Tense and sometimes bitter political wrangling had surrounded its acceptance. Still to be faced were nine years of engineering challenges in design, construction, and testing before it was ready for launch. After that, eight long years were required of its protagonists to construct history's greatest star map from the satellite survey.

M. Perryman, *The Making of History's Greatest Star Map*, Astronomers' Universe
DOI 10.1007/978-3-642-11602-5_2, © Springer-Verlag Berlin Heidelberg 2010

Why should the ancient and seemingly abstract scientific niche of pinpointing the positions of the stars have ranked so highly amongst the many ideas fiercely competing within the European science communities for space experiments? Why should Europe elect to invest hundreds of millions of euros in this seemingly arcane enterprise? Why should established scientists choose to devote decades of their lives to improving our knowledge of the positions of the stars?

The motivation—as for all space missions—involved issues of national and international prestige, the opportunity to develop cutting-edge technical capabilities across Europe, and pure scientific enquiry, not necessarily in this order, but to some extent in equal measure.

TO UNDERSTAND WHY THE MEASUREMENT of star positions was being pursued at all, present-day knowledge of our place in the cosmos forms an indispensable backdrop: we need some understanding of the nature of the stars, how they form the building blocks of galaxies, and how galaxies point to the way that the Universe began.

It might sound surprising that star positions provide an important foundation across all of astronomy. The reasoning is broadly as follows: careful measurement of their positions allows us to determine their distances by geometric triangulation. Once we know star distances, we have direct access to their basic physical properties such as their energy output and their mass. This is information enough, with some theory and some computer modeling thrown in, to figure out their internal composition, their ages, and the history of their nuclear-fuel burning. We can trace back from their fossil narrative their origins and their past evolution, and use these models to predict their future destiny.

At the same time we can actually watch how the stars are moving through space. Tracing their paths on the sky gives strong clues as to how our Galaxy[2] is assembled and, more surprisingly, points back to how it came into being in the far distant past. Measuring star positions, in short, provides a rich stellar census telling us much about the community of stars around us. All of these clues also allow us to piece together a detailed picture of our own Sun's past history, and our own place within our Galaxy and the Universe as a whole.

The progress of the field over the last two thousand years brings home the complexities of the Universe in which we are immersed. Major scientific breakthroughs were made as the measurement of star positions improved over centuries past, and it was profoundly evident that further advances in our knowledge of astronomy could be guaranteed if much more accurate star positions could be measured.

But trying to pinpoint their positions from the ground had hit an insurmountable barrier imposed by Nature: our Earth's atmosphere. Science was confronted by a range of questions which were clearly posed, but which could not be answered using Earth-based telescopes, however large. Observations from space would give a far clearer, crisper view of the stars. A mapping of our Galaxy from high above the atmosphere could be guaranteed to push our understanding of the Universe to new limits.

THE NIGHT SKY IS POPULATED by stars. Several thousands are visible to the naked eye from a dark location, and countless millions more are revealed by the largest telescopes on ground. Since the 1990s, and the launch of the Hubble Space Telescope, access to space has given us views across our Galaxy and beyond with even greater acuity.

Present-day light pollution aside, our distant forebears had much the same view of the night sky. They had no idea what the stars were, and this state of happy ignorance persisted until rather recently. Our ancestors would have stared at these sentinels of the night, if not in disinterest, most certainly in incomprehension—at least from the time that *Homo sapiens sapiens* acquired an ability for abstract thought and an elementary understanding of the physical world, perhaps some hundred thousand years ago. Thereafter, over countless aeons, nothing much changed in their understanding of the heavens. The stars were invested with mythical meaning, but ignorance of their deeper nature was immutable over many thousands of generations.

One early view held, rather poetically in hindsight, that the stars represented tiny holes in a celestial vault revealing miniscule glimpses into a source of great brightness beyond. Whether the heavens were of fire, whether there was an infinite void beyond the stars that rotated, or whether the stars shone by light reflected from the Sun, all were debated keenly by the ancient Greeks. Yet even our modern factual ideas, which are taken so much for granted by so many, are still not universally shared: I saw at first hand how many people remained inside for the total solar eclipse which plunged the Ghanaian capital Accra into daylight darkness in 2006, a stunning spectacle of nature still accompanied by mysticism and fear. And I sat one night in the silent Saharan darkness, on sand dunes outside a small nomadic village beyond Chinguetti in Mauritania, over which another total eclipse had made its ghostly traverse in 1976. There, with the star-studded heavens bearing down startlingly upon us, I was solemnly assured that shooting stars signify departing human souls.

It is easy to dismiss such ideas quizzically. But our current scientific views have only been earned with the insight and dedication of past generations of thinkers. We inherit their gifted progress, and usually take it for granted. Peering out from stone-age shelters millennia ago, we too might have been hard pressed to figure out the internal nature of the stars and the early history of the Universe. We see so far today by "standing on the shoulders of giants", a delightful metaphor oft parroted in science. It is an aphorism attributed to the twelfth century scholar Bernard of Chartres, used more famously by Isaac Newton, and stamped into our consciousness on the edge of the standard British two pound coin to remind us daily lest we forget.

BEFORE CONTINUING WITH OUR current views, let me explain the process by which scientific knowledge in general, and astronomical knowledge in particular, advances at the turn of this millennium. How do scientists dare to be so confident that the present theories and paradigms that underpin their advances are true?

Founded in 1919, the International Astronomical Union lists in its membership around ten thousand professional astronomers in eighty six countries. Thousands of new scientific papers are published each year in recognised scholarly journals, each reviewed by independent peers for originality, authority, clarity, and credibility. I give these numbers to illustrate that astronomy is a reasonably large and highly active business.

The career of a professional scientist rests on coming up with original ideas or new experimental results. These may be important in confirming hypotheses previously published or, of equal value to science, proving beyond reasonable doubt that previous ideas or analyses are false. In this almost Darwinian manner, correct ideas become progressively more consolidated, while those falsified are dropped from further consideration. Science sets as its goals the objectives of knowing and understanding the natural world, by following a course that is wholly dependent on empirical evidence and testable explanations.

We live in a world in which society supports scientific research, and in which governments fund research institutes and university departments to teach and to stimulate enquiring minds. Aspiring young scientists compete to join existing research groups, and thereafter contribute their own insight, building and nurturing their own creative reputation. At the very basic level, theories are publishable if they explain hitherto unexplained results, or if they make plausible new predictions that might be tested experimentally. Individuals or groups around the world may compete, or collaborate, to develop new theories or point to weaknesses or errors in existing ideas. They may petition for funding to build new instruments that are expected to yield more advanced results that may challenge existing theories, or demand new ones. Big facilities might demand years of lobbying, and a decade or more to design and build.

An important discovery draws in others to join the excitement, and the hope of refining or extending the results. Competition acts as a stimulus, and explains why science is self-correcting. Ultimately, new experimental results are acquired, theories with considerable explanatory or predictive power gain favour, as do the scientists whose insights and advances are widely acclaimed. Inadequate explanations are discarded. Ideas based on no substantive reasoning are ignored.

I should stress that, in the language of science, a theory is far more than a simple hypothesis, an unproven guess, or a deeply-held belief. To be generally accepted by the scientific community as a whole, a theory must have been verified multiple times by independent researchers, and must be capable of making predictions which can be verified by specific observation. But, even so, no theories are guaranteed to last. Like coconuts at a fairground shy, even the most established stand as a temptation for the suitably competent to try to knock down. Occasionally, even the most revered theory can be dismantled, guaranteeing recognition for whoever succeeds in so doing. And an entire field of science can undergo a revolution following a pioneering breakthrough or profound insight.

Piece by piece, occasionally through a major discovery or revolutionary theory, but more typically through many modest increments, comprehension advances. And so we arrive at the present inherited state of knowledge which we dare to describe with such confidence.

EACH STAR IS, we now infer, a gigantic sphere of gas, predominantly hydrogen, held together by its own immense gravity, squeezed to high densities and temperatures by this overpowering force, and releasing vast amounts of energy through nuclear burning.

Our own Sun is just one such star amongst a seeming infinity of others; large, bright and special to us because of its proximity, and the centre of our solar system. Our Sun is a hundred times the diameter and a million times more massive than the Earth, and nearly five billion years old[3]. It appears so enormously much brighter than the other suns ranged across the night sky, bathing Earth in radiated daylight, simply because it is so very much closer. Only the light of the Sun reflecting off other bodies in the solar system—the planets and our Moon—makes them shine, for they have no light of their own. Similarly, everything we see in the world around us, we see by the light of the Sun, light which was generated deep in its interior.

The energy source of the Sun and all other stars, nuclear fusion, was understood convincingly only in the mid-1900s. The interlocking edifice of physics and mathematics necessary to explain it represented a triumphal advance in twentieth century science.

Decades of intensive astronomical research have since shown that stars exist in a truly bewildering variety: of sizes and masses, ranging from very much smaller than our Sun to very much larger; of temperatures, ranging from much cooler than our Sun to very much hotter; and of ages, ranging from billions of years younger to billions of years older. We recognise very different chemical compositions too: ranging from almost pure hydrogen, to stars much more polluted by heavier elements produced from their own nuclear burning, or born from chemically-rich debris of previous generations of stars long gone. Not only can we measure these widely-varying properties, but we now have compelling models to explain their origin.

Wikimedia Commons

$E = mc^2$

Energy production within the stars is prodigious in the extreme. Our Sun consumes its hydrogen fuel at a bewildering five million tonnes a second, a gas guzzler par excellence. It beats hands down the consumption of a large coal-fired power station, of around one hundred kilograms a second. This monstrous mass is lost from the Sun forever, carried away into space in the form of heat and light, and some sub-atomic particles. We can try to imagine the Sun's mass lost in this way as a cube of material, one hundred meters on a side, vanishing each and every second of its almost timeless existence.

What we cannot grasp is the unimaginable energy released as this colossal mass vapourises as a result of nuclear fusion. For each *gram* converted into pure energy, equivalent to the weight of a typical bank note, Einstein's mass–energy equivalence principle (the famous $E = mc^2$, connecting energy, mass, and the speed of light) tells us that twenty five million kilowatt-hours of energy emerges to take its place. Just one gram of fusion power gives the same amount generated by the largest coal-fired power stations in a day, enough to keep a sizable fraction of the population of Europe or the USA in electricity for around an hour. It is why the grail of nuclear fusion is such a coveted energy source.

Even with this stupendous loss of mass each second, stars continue to shine for billions of years because their reservoirs of nuclear fuel are so unimaginably immense. Thankfully, Nature has also cleverly arranged that not all of the fuel is heaped onto the conflagration at the same time. Rather, complex processes of self-regulation prevent stars from catastrophic runaway burning and collapse. As the interior heats, more energy is released, and this energy pressure pushes material outwards which prevents further collapse. In this way the ball of awesome fire is maintained in a delicate state of overall equilibrium.

Theories of nuclear fusion, and detailed models of star interiors, are now remarkably advanced. Our knowledge of physics describes these processes rather precisely. We have great confidence in their explanatory and predictive powers, such that we can state with conviction that the Sun came into existence a little less than five billion years ago, collapsing from a region of dense gas clouds; and that it will continue to burn more-or-less in its present form for another five billion years.

Close-up view of our Sun's surface

These numbers tie neatly together with the geological records on Earth, the ages of meteorites, models of our solar system's formation, observations of other stars, and a whole host of other results. Astronomy, in short, provides us with very clear guidance as to how our Sun has evolved over the vast sweeps of geological time past, and how it will slowly change over the almost timeless eternity of the future. Society has many pressing concerns as we enter the third millennium, but whether the Sun will continue to rise each morning—a preoccupation of antiquity—is not one of them.

OUR GALAXY IS JUST ONE amongst a seeming infinity of others, special to us because we reside within it, but comprising billions of stars. They are assembled in a diversity of curious ways. Some of the constituent stars appear to travel through our Galaxy quite alone. Others turn around each other, as binary twins, sometimes even as triplets or quadruplets. Some binaries circle each other so closely, their surfaces almost touching, that their orbit periods are as short as days or even hours. Others rotate about each other at monstrous separations, at the very limit of their feeble gravitational attraction, taking many thousands of years to complete a single orbit.

Many young stars reside in larger localised groups or clusters, starting their lives with different masses but sharing the same chemical properties, inheriting their elemental composition from the gas clouds out of which they formed. Within our Galaxy, hundreds of known star clusters mark the ancient debris of stellar nurseries. Those nearest to us, the Hyades and the Pleiades, are concentrations of stars visible by eye from a dark location. Some older star clusters, comprising hundreds or even thousands of stars, have survived in beautiful formation for tens of millions of years after their birth, held together by their own gravity field. They oscillate gracefully about each other in an all but perpetual dance, albeit not at a rate that can be discerned, not even over a human lifetime.

The Milky Way, seen from Death Valley

While most of these denizens of our Galaxy are visible only through telescopes, and their detailed nature is revealed only by careful study and analysis, one dominant assembly of stars is clearly visible to the naked eye on a clear dark night: the great concentration known as the Milky Way, the luminous irregular band of light stretching across the night sky. The Milky Way is the name given to the flattened disk of our own Galaxy, a massive pancake-shaped agglomeration of hundreds of billions of stars, all in a state of steady rotation around its central axis like a slowly-turning wheel. Superimposed within this disk-like structure, a significant fraction of the stars are actually laid out in a vast multi-armed spiral form.

Our Sun is just one rather typical star, far from our Galaxy's centre. As Bertrand Russell so eloquently put it[4] "Life is a brief, small, and transitory phenomenon in an obscure corner, not at all the sort of thing one would make a fuss about if one were not personally concerned."

Presiding over our own solar system with its planets, moons and asteroids, our Sun lies so far from the Galactic centre that light takes tens of thousands of years to cross this vast gulf of space. We are so far out that the Sun takes a significant fraction of even geological time, two hundred and fifty million years, to make a full circuit around our Galaxy. Just as our Earth is trapped by gravity in its indefinite circling of the Sun, so our Sun endlessly orbits the Galactic centre, held captive by its overall gravity field. Constructed on a truly staggering scale, no more than forty such rotations of our Galaxy have been swept out since time itself began.

From Earth, in orbit around the Sun, we can stare out in different directions across the vast lonely expanses of the Galaxy in which we are immersed. Astronomy tries to figure out the structure of the billions of stars which make up our island universe, and to comprehend how it all came into existence.

Even our own imposing spiral Galaxy, the vast Milky Way, whose size all but paralyses the imagination, is but the start of any imaginary voyage across the visible Universe. Our largest telescopes reveal preposterously larger gulfs of empty space beyond our own Galaxy, before arriving at any others. From the southern hemisphere, from which the central region of our Milky Way Galaxy can be studied, we can even see our nearest neighbouring galaxies with the naked eye: the Magellanic Clouds, Large and Small, appear as smudges of diffuse light, the feeble glow of their billions of stars viewed across a large fraction of eternity. These, and other nebulae in our local galaxy grouping which spring into view through the telescope, precede ever larger voids of nothingness. We must cross these pointlessly empty spaces, across which light itself takes millions of years to travel, before we encounter other groups of galaxies.

This sequence of all but meaningless voids and galaxy clusters extends ever further, as far as we can tell, on and on, perhaps into infinity. From the largest telescopes on Earth, hundreds of billions of galaxies, some seemingly like our own, some evidently very different, stretch out in all directions. Each contains hundreds of billions of stars, of which our Sun is just one.

IF ALL THIS IS NOT INCOMPREHENSIBLE enough, the picture gets yet more curious. Since the early 1900s and the pioneering deep sky observations of Edwin Hubble, we know that all of the galaxies beyond our own are receding from us at ever-increasing speeds the further out we probe. Our Universe, a term used to embrace all of space and all of time, everything which we can possibly observe, is believed to be relentlessly expanding, flying apart.

We can trace these galaxy motions backwards in time, with a strange result. The expanding Universe seems to have originated a little more than thirteen billion years ago when, as far as we can tell, everything now contained within it emerged from a single point. This moment is taken as representing the very origin of space and time. This is, admittedly, more than just a shade perplexing. But how we think the Universe might have come into existence is very much more puzzling still.

The theory of quantum mechanics is an essential ingredient in explaining current thinking. This is a baffling description of the microscopic world, couched in difficult mathematics and abstruse probabilities. In the quantum world, for example, an electron no longer occupies a given place at a given time, but exists as if it is everywhere and nowhere. This vague-sounding claim can actually be cast into a formal mathematical framework, explaining and predicting a whole host of observations that cannot be accounted for by classical physics. Its wide application in modern technology shows that it has important practical uses too. So, mysterious and perplexing though it appears, as much as it defies our intuition and logical interpretation, physicists view it as a true theory of the sub-atomic world.

So far so good, but here is the problem. According to the uncanny prescripts of quantum mechanics, energy can be created, spontaneously, from a complete vacuum. Tiny amounts can spring into existence, as if from nothing, just so long as it agrees to disappear again promptly, within a fraction of a second. We have good evidence for this occurring in the world around us, too. Dragged mentally kicking and screaming, we have little choice but to accept this perplexing fact: matter can appear where none was before.

And for my next trick...

Julia Kostelnyk

But in an extreme and preposterous manifestation of this theory, equal amounts of positive energy in the form of matter, and negative energy in the form of a massive gravitational field, appear to have sprung into existence from nothing at all. Our entire Universe, all the galaxies and all of their constituent stars, emerged out of nought.

Having appeared from nothing and from nowhere, another extraordinary surprise awaits us. For so finely balanced was this sleight-of-hand, mass on one side of the ethereal scales, gravity balancing out the other, that the total energy in the Universe summed to precisely zero. Under these finely tuned circumstances, the laws of quantum uncertainty permit this pleasing state of affairs—this spontaneous borrowing of matter where absolutely nothing existed before—to persist forever. This supremely incomprehensible event is referred to, with almost nonchalant confidence, as the Big Bang. It is a label which normalises the fantastic.

From a number of lines of evidence, mainly from the expansion of galaxies and faint immortal echoes of the resulting ripples in space–time seen in the cosmic background of microwaves, it is timed to have occurred a little more than thirteen billion years ago. This is, let us stress, an awesome passage of time. So utterly remote in the past, a patient traveller could cross the entire United States in the time elapsed since the Big Bang by taking just one leisurely step every five thousand years[5]. The momentous event sits centre stage in modern cosmology, the pinnacle of our efforts to study the Universe on its largest scales of time and space.

Indigestible though this extraordinary picture might appear, the uncomfortable reality is that all of the evidence that we have today in science—one vast interlocking edifice of observational data and detailed theories from astronomy, physics and mathematics—all point towards it being largely correct. For the majority of astronomers Big Bang cosmology forms the theoretical paradigm, the foundation of our understanding of the Universe. It has made sweeping predictions, and has generally passed all tests thrown at it with flying colours. It is, accordingly, far more than a simple hypothesis, although it is still not accorded the stature of a scientific law, or unquestioned axiom. It may seem bizarre, and it may yet be proven wrong, but it remains the most serious contender for explaining the emergence, if not the ultimate origin, of our Universe.

Yet our knowledge, and our confidence, shudders to a halt when we confront the still grander questions that this picture raises. Where or what was the nothingness that our Universe emerged from? Did nothing at all exist before the Big Bang, not even time itself? Why do matter, time, and space have the properties they do? And the big one looming above them all: *why* does the Universe exist? Here we can at least be unequivocal: science simply offers no idea.

REMARKABLY, THOUGH, and however inadequate our understanding of the Big Bang itself, scientific progress over the last few decades has provided an impressively detailed and consistent picture of what happened in the seconds, minutes, years, and thousands, millions, and billions of years after time zero. Our current understanding of how the Universe evolved after the Big Bang has been told many times in recent popular literature, and the essence is as follows.

Very broadly, we believe that the lightest elements came into existence within the first few minutes following the Big Bang, as the matter which had been created spontaneously expanded dramatically and cooled rapidly. Galaxies grew, controlled by the force of gravity, from tiny density and pressure ripples which somehow appeared and reverberated in the early Universe. Stars then formed from collapsing and cooling clouds of hydrogen gas, eventually forming the staggering number and strange diversity of galaxies that we see today.

Within the first-born galaxies, the most massive stars consumed their primordial nuclear fuel rapidly. Deep within their nuclear furnaces they created new elements heavier than hydrogen and helium. At the end of their lives, their celestial ash was dispersed, flung far into space through supernovae explosions and other outflows of dying stars. Out of this elemental soup, new generations of stars would duly form, shine for millions or billions of years, and expire in their turn as their fuel ran short. On and on this mighty play will be enacted, until the Universe itself runs cold. Until then, as nicely summarised in Ecclesiastes "there is no new thing under the sun."

In the last few years, the recipe for our Universe's birth has demanded some new ingredients. In the process, it has become still more perplexing, considerably more confusing, and somewhat more hypothetical. One of the favoured constituents is an early period of enormous 'inflation', in which an embryonic pea-sized Universe—we are asked to take this size more or less literally—expanded outwards at a staggering rate. According to current ideas, this inflationary force provided a massive outward push which prevented its intense primordial gravitational field from collapsing back on itself immediately after its formation.

Mysterious dark matter and, more recently still, dark energy, have been proposed as yet other peculiar constituents of this remarkable story. Complex mathematical ideas are being put forward as explanations, with exotic names such as vacuum fluctuations, super-symmetry, string theories, chaotic inflation and quintessence. But I hasten over these troublesome concepts because they are not central to our own story, although they provide the backdrop as to how cosmologists today believe the Universe began.

Yet history shows only too well how the supreme confidence in scientific understanding of the Universe of one generation may in turn be swept aside by future authorities, perhaps to be replaced by their own perplexity at the feeble comprehension pervading at the time. Such misplaced confidences abound throughout history, and I give just two pertinent examples.

Archbishop James Ussher (1581–1656) is renowned for deriving a scholarly chronology that placed creation on the night preceding 23 October 4004 BCE[6]. The Big Bang, our current scientific paradigm of the origin of the Universe, has extended its age a shade, and now confidently asserts that everything came into being from nothing a little more than thirteen billion years ago, the elements taking but a further few minutes to cook.

Library of Congress: Bain Collection

Arthur Eddington

Much later, in 1932, Sir Arthur Eddington, Plumian Professor of Astronomy in Cambridge, declared that the radius of the Universe, before it began to expand, was a little more than a billion light-years[7]—a far cry from the pea-size cosmos we now try to contemplate, but averred with equal confidence. Eddington admitted that his figure was unlikely to be the final word, but thought that it was "not likely to be in error by more than a factor of two."

Will the Big Bang, and its attendant tweaks, pass future observational tests, and duly become fully embraced as theory, scientists' ultimate achievement in comprehending and describing the origin of the Universe? Or, will its latest bells and whistles more closely parallel the contrived epicyclic theory invoked by the ancient Greeks to explain planetary motions which reigned unquestioned for a thousand years before it perished? Will it eventually prove to be nothing more than a chimera, the latest of the emperor's new clothes to be found abandoned in the astronomers' wardrobe? While most are swept along by the euphoria of its successes, the sceptic might still reasonably ask "come now, everything emerged from nothing—and you expect me to believe that?" Yet either everything was created from nothing, or everything has always been there. Which of these indigestibles do you prefer? Only future scientific enquiry may tell.

WHATEVER OUR VIEWS of the Big Bang, the present state of the Universe is laid out before us, for enquiring minds to inspect and ponder. It certainly throws mystery upon mystery at us. The more one digs into physics and delves into astronomy, the more one cannot help but be impressed and confused by the perfect conditions and perplexing coincidences which seem to be piled high, one upon the other, so as to allow our Universe to exist in a form suitable for life: the fine balancing of forces that makes the building blocks of everything look the way they do, and the elements with their plethora of astounding properties which allow so many other chemicals to form. We can grapple with the immense age of the Universe needed to give life time to develop, and struggle with the unimaginable number of stars allowing small probabilities of habitability and evolution to work their cumulative magic. We can reflect on the almost endlessly perfect suitability of our own solar system for life. In all of these, and countless others, the coincidences seem almost too great.

Winston Churchill, who famously viewed Russia as "a riddle, wrapped in a mystery, inside an enigma", would surely have added a few more matryoshkas had he studied the cosmos.

Many in science are actively debating the dilemma that all of these many baffling coincidences pose. Is the Universe fine-tuned such that life can only occur when certain universal physical constants lie within a very narrow range? Are we part of a 'multiverse', comprising multiple universes each with different physical constants in which we are, by definition, only able to inhabit a hospitable one? Must science somehow take into account the anthropic principle, in which the only kind of universe that humans *can* occupy is one that must be similar to our own? Does this itself in turn imply an intelligent designer, or does such a teleology undermine any search for a deeper physical understanding of Nature? Does our Universe *have* to be so big and so complex, simply to allow life to exist perhaps just once throughout all of space and all of time?

WE NOW LEAVE THESE GRAND QUESTIONS to one side, having set out the context. This story focuses on what happened long after the Big Bang dominated the headlines. Once the primordial conditions of the exploding Universe had been enacted, once the physical constants ruling matter had somehow been defined, all of existence was set on its inexorable path into eternity.

Galaxies were created, and stars within them, not in some static and immutable form, but as building blocks of an entire Universe in perpetual turmoil, ceaselessly interacting. In this constantly restless panorama, galaxies are collapsing, cannibalising and merging with their punier neighbours, or being torn apart as they venture too close to a greater colossus. Meanwhile, the myriad stars within them are being born and shine for aeons, each following very different lives and very different deaths depending on their size and their chemical heritage, all the time swirling in complex and multiply interacting patterns within their parent galaxies.

Many stars, we know now, have bred planets encircling them as they settled down into their own fleeting semblance of permanence. At least one planet in the uncountable infinity which might exist has been endowed with enquiring minds which now look outwards, trying to interpret the almost uninterpretable. This is the stage on which our own play is set.

Why Star Positions?

Two stars keep not their motion in one sphere.

William Shakespeare, King Henry the Fourth, Part I

THE ENTIRE *raison d'être* for the Hipparcos satellite, and its resulting celestial cartography, was the accurate measurement of tiny angles that divide up the sky. In order to understand these measurements, some essential concepts must be laid out.

Dividing a circle, whether on paper or on an imaginary sweep of the celestial sky, is a task well-posed in principle. Practical techniques for doing so aside, it is only necessary to agree on the unit of subdivision. [Scientific users today prefer to work and calculate angles in units of radians. In this natural mathematical system, a full circle is divided into $2 \times \pi$ radians, where π or 'pi' = 3.14159 is that most remarkable of numbers relating the circumference of a circle to its diameter. Radians are wonderfully convenient for computations, but not very intuitive, and I won't mention them again.]

The commonly accepted choice of three hundred and sixty degrees in a circle was made for us long ago. Ascribed to the Sumerians of ancient Babylonia, more than 2000 BCE, it was perhaps guided by the number of days in a year. One degree was subdivided into sixty minutes of arc, and each minute of arc was divided still further into sixty seconds of arc. As a result, one degree of angle comprises precisely three thousand six hundred seconds of arc (60×60), and one complete circle therefore contains a little more than one million of these tiny seconds of arc $(360 \times 60 \times 60)$. The curious choice of sixty rests on the number itself being highly composite: it has many divisors, which facilitated calculations with fractions performed by hand.

M. Perryman, *The Making of History's Greatest Star Map*, Astronomers' Universe
DOI 10.1007/978-3-642-11602-5_3, © Springer-Verlag Berlin Heidelberg 2010

These subdivisions into sixty are the same as those used to carve up and so describe the passage of time. We still happily cling to our day divided into hours, minutes, and seconds, the only common measure to have firmly resisted metrication. These byzantine subdivisions are, to be sure, terribly cumbersome for modern scientific use and computer manipulation, as inconvenient to calculate with as the British pre-decimal currency of twelve pennies in a shilling, and twenty shillings in a pound. But the awkward system of degrees, minutes and seconds remains in widespread use today. It is the system still taught at school for measuring angles, and it is similarly imposed on us in the representation of geographical coordinates of latitude and longitude on Earth.

The same system is used in astronomy to describe the positions of objects on the sky. Astronomers use the terms right ascension and declination on the sky to correspond with longitude and latitude on Earth, although we need not touch on these again. Referring to 'minutes of arc' and 'seconds of arc' simply reminds us that we are dealing with angular measurements, rather than minutes or seconds of time. We will keep our description to angles measured in degrees, minutes, and seconds, and give some examples to try to visualise their magnitude.

Luc Viatour

Solar eclipse of 1999

If we could see the entire sky—which we can't because the Earth's surface on which we stand obstructs around half of it at any time—a full sweep around the celestial vault would correspond to 360 degrees. The Sun and the Moon both cover the same *angle* on the sky, about half a degree. Although their angular size is coincidentally the same, these two celestial objects are of very different physical sizes—the Sun being a massive 1.4 million kilometers in diameter and our Moon being only 3500 kilometers. Hold out your thumb at arm's length and it covers about the same angle, even though it's much smaller still. An angle is simply the ratio between size and distance. The Sun is larger and much further away than the Moon, and it is by mere coincidence that the two occupy roughly the same angle. This is why, when the Moon happens to pass across the face of the Sun, it all but blocks it out completely, leaving that miracle of the natural world, the hauntingly spectacular total solar eclipse.

As a small aside, the rarity of this celestial performance comes down to the fact that the Moon orbits the Earth in a slightly different plane to that of the Earth orbiting around the Sun. As a result, the simultaneous and perfect alignment of all three does occur from time to time, but is rather uncommon.

If the Sun ventures a little closer than normal, or the Moon a little further, which happens because the orbits are slight ellipses and not perfect circles, then an 'annular' or ring-shaped eclipse is the result. The ability to predict eclipses is possible because accurate positions of the various bodies in the solar system observed over many decades have been tied together through a detailed model describing their gravitational interactions. Predictions well into the future allow eclipse chasers to plan the next spectacle with confidence. Armed with predictions over the distant past, historians can date certain documented events very precisely, from which other dates, and even a society's calendar, have been reconstructed.

SO WE CAN CERTAINLY VISUALISE half a degree as the angular size of the Sun or Moon. But how about the much smaller one second of arc? Imagine people standing in a large circle, spaced one meter apart. The circle would have to be two hundred kilometers in radius, the distance between London and Sheffield, or a little less than New York to Washington, or Munich to Vienna, for each person to be spaced by an *angle* of one second of arc when viewed from the centre. A million volunteers would be needed to populate the complete circle.

This very small angle, one second of arc, turns out to be a rather convenient angular measure in astronomy, and it has been used to construct a very basic measure of astronomical distances, the parsec. This is central to the story, and I will return to it shortly.

In very round numbers, one second of arc is also the angle to which astronomers can measure, with relative ease, the position of a star at any one moment. Better accuracy than this is not at all easy, and the limit arises due to the shimmering effects of the Earth's atmo-

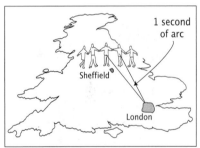

An angle of one second of arc
(greatly magnified)

sphere. It is the atmosphere which pushed these measurements to space, and a little background is useful to explain more carefully why.

The lowest portion of our atmosphere is known as the troposphere. Extending to a height of about ten kilometers, it contains three quarters of the atmosphere's mass, and almost all of its water vapour. The word is handed down to us from the Greek 'tropos' for 'turning' or 'mixing', and eloquently conveys the fact that turbulent mixing of the air, due to convective heating rising from the Earth's surface, plays an important part in our atmosphere's structure and behaviour. Most phenomena associated with day-to-day weather also occur in this region. Air gets thinner all the way up due to the diminishing effects of gravity, and it doesn't suddenly disappear at the top of the troposphere. But the balance of heating and cooling changes, and turbulence decreases—which is why commercial airlines prefer to fly high in the troposphere. Higher still, in the stratosphere, the turbulence all but disappears. The Earth's atmosphere only gradually merges into interplanetary space, and the boundary is variously taken at somewhere between a hundred and ten thousand kilometers.

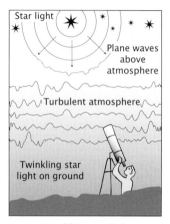

Twinkling star light

Turbulence affects light rays passing through the atmosphere, and causes the familiar twinkling of star light. Eratosthenes (276 BCE–194 BCE) commented on their 'tremulous motion'. Turbulence increases above a hot road surface in summer, or around evaporating petrol fumes in a filling station forecourt, giving an even more exaggerated rippling motion. To picture the problems of measuring star positions from the ground, imagine trying to pin-point an object seen through such a shimmering haze. To minimise the effects, astronomers build their telescopes at high mountain sites where the thinner atmosphere and smaller turbulence gives more stable images, and away from the bright lights of roads and cities. At good sites, the dancing motion might drop below a second of arc, but not by much more.

The atmosphere is crucial for life, but it makes things hard for astronomers, for its rippling motion is the main obstruction when it comes to getting better astronomical images from the ground. The Hubble Space Telescope has achieved its spectacular breakthroughs by riding high above the Earth's surface, beyond the atmosphere, at a height where this complicated image motion is non-existent.

The same twinkling motion of stars viewed through the atmosphere has imposed an impenetrable barrier to measuring their positions to much better than about one second of arc. And for the very same reason, the Hipparcos satellite was eventually able to achieve its impressively accurate celestial cartography by conducting its survey of the night sky from high above the Earth's atmosphere.

THE HUMAN EYE imposes its own limit to measuring angles of about one minute of arc—you can see two distinct car headlights at night up to about a kilometer distance, but much beyond that they merge into one. Mainly determined by the small diameter of the pupil through which light enters the eye, this limit is many times worse than that inflicted by the atmosphere. Until the invention of the telescope, observations by eye had therefore placed even worse limits on the accuracy of star positions. The introduction of the telescope, credited to Dutch opticians in the opening years of the seventeenth century, but considerably and more famously improved upon by Galileo in 1609[8], brought with it two distinct improvements. First, it extended the faintness limit of the stars that could be seen, revealing countless more than were visible by eye. The larger diameter of the telescope aperture also gave an improved accuracy of positional measures.

Making the telescope mirror larger improves the accuracy proportionately, but only up to the point that the atmospheric shimmering motion sets in at around one second of arc. After that, even very large telescopes on the ground, the largest now reaching a diameter of ten meters, fail to break through the accuracy limit on angular positions set by the listless atmosphere.

Europe, the USA, and Japan, all now operate big telescopes of around this size. But ambitious plans are currently being drawn up on both sides of the Atlantic in an unspoken race to construct the first telescope mirrors of gargantuan diameters, thirty or forty meters in size. Their protective domes will be the diameter of the Royal Albert Hall in London but twice as high. With such a large area of glass they will see much fainter objects, and probe

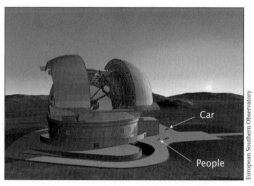

Europe's Extremely Large Telescope concept

further than ever before. But if used to measure star positions, even they will not be immune from the intractable flickering motion of the atmosphere—and its impenetrable accuracy barrier.

SO MUCH FOR ONE SECOND OF ARC. It is a tiny angle, problematic enough to measure and, as we will see later, it proved a great challenge for the astronomical instrument makers of earlier centuries. But in the world of Hipparcos it's an angle that is left far behind. For the accuracy of the angular measurements that the satellite targeted was one thousandth of one second of arc, a thousand times better than the jittering motion seen by telescopes on the highest mountain sites. The entire celestial sphere, and more than a hundred thousand stars distributed across it, were pin-pointed to this exquisite accuracy.

One thousandth of a second of arc, viewed from Europe

In the picture of our circle of reluctant volunteers spaced one meter apart, the circle would have to grow to a radius of two hundred thousand kilometers for people around its rim to be separated by this angle of one thousandth of a second of arc when viewed from the centre. If this fails to properly portray such a microscopic angle, we can try to imagine it as the size of an astronaut on the Moon viewed from Earth, a golf ball in New York viewed from London, or the diameter of human hair seen from ten kilometers away. Or try this: stand one meter away from someone, and focus on the end of one hair on their head. Hair grows at about one centimeter a month, so that in one second when viewed from this distance, their hair has lengthened by an angle of one thousandth of a second of arc.

Building a network of star positions across the sky with this accuracy represented a technical challenge of enormous delicacy. The effects of the atmosphere and Earth's gravity had to be banished, mirrors had to polished, instrument temperatures had to be controlled, and structures had to be stabilised in a manner never before achieved. But doing so allowed the construction of a celestial star map whose remarkable precision changed the panorama of astronomy.

WHY DOES MEASURING THE POSITIONS of stars offer any insight at all into their properties, let alone the structure of our Galaxy, or the origin of the Universe?

The explanation is in fact not strictly related to pinning down the position of a star on the sky with great accuracy for its own sake; measuring an accurate position as such is not the ultimate objective. Rather, the positions of stars in the sky vary minutely over months and years for a number of reasons which are central to the task at hand. The crucial point is that repeatedly measuring an accurate position over a period of months and years can discern and track certain tiny motions which prove central to understanding their nature.

So, what are these motions? Let me start by explaining what they are not. Those with a little familiarity with the night sky will know that the stars do not appear in the same position from night-to-night, nor even between the start of the night and the end, but *appear* to move slowly across the night sky. To all intents and purposes, at least until we start our exploration of the heavens as measured by Hipparcos, the stars occupy fixed positions relative to each other, and it is simply the spinning of the Earth on its axis, one rotation every twenty four hours, combined with its motion around the Sun, once per year, which together give the stars their apparent collective movement. At least as far as the human eye is concerned, their relative positions are preserved without change over hundreds of years. This apparent systematic rotation of the heavens, an artifact of the rotating Earth, is not the effect that Hipparcos set out to measure.

EVERY STAR IS MOVING THROUGH SPACE. We know this now, although three hundred years ago humanity did not. As a result, over many decades or centuries, small displacements of some of the most swiftly moving stars do begin to be discernible. The manifestation of these star motions was first reported by Edmond Halley in 1718.

Halley, of his eponymous comet fame, had been comparing his contemporary star charts, still constructed with relatively coarse instruments with position accuracies of only around ten seconds of arc, with those made and recorded by the ancient Greeks. He noted that the bright stars Aldebaran, Arcturus and Sirius had moved significantly from their positions given by Ptolemy in his great mathematical and astronomical treatise, the Almagest. Sirius, for example, had moved nearly half a degree southwards, about the diameter of the Moon, over the intervening two thousand years[9].

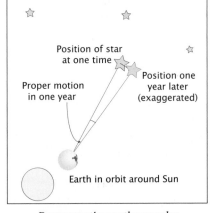

Proper motion as the angular change of a star's space motion

Although their movements across the sky over a year or so were very small, their progressive motions had accumulated over time. His contemporary catalogues, accordingly, charted star positions which were significantly displaced from those originally seen centuries before. It must have been a discovery of enormous excitement for Halley, and the first detection of star motions across the sky was a turning point in the studies of the stars. Many other shifts were soon reported, and the collective study of stellar motions was born.

If the motion of any one star in the sky could be examined in detail, greatly magnified, each star would be seen to move in a straight line across the sky. The pattern of motions might appear rather chaotic, in the sense that even stars near to each other on the sky can be moving in very different directions, and at very different speeds.

Measuring how the angular position of each star on the celestial sphere changes with time gives what astronomers call the star's 'proper motion'. The name is a little cryptic, probably drawn from the French 'propre' for 'own', but was used to make clear that what is being measured is the motion originating from the star itself, and not that due to other effects like the Earth's rotation. It also reminds us that what has been measured is an angular shift over time, and not the actual speed of the star through space.

The two are closely related, but the distinction is important. What we see is simply the star's movement projected onto the celestial sphere, which we can only describe in terms of an angular motion. If we don't know the star's distance, we cannot infer its true motion through space: a star whose position has changed by a certain amount over a few years might be a relatively nearby star moving slowly through space, or a star at a greater distance moving more rapidly. In practice, stars with large proper motions do tend to be nearby, and searching for high proper motion objects—by comparing photographs of the sky taken a few years apart—has proved to be a bountiful way of discovering nearby stars. Most stars, meanwhile, are far enough away that their angular proper motions are very small.

Velocity is unknown if an object's distance is unknown

The star's space velocity is an important quantity for astronomers, but determining it from the angular motion needs a knowledge of the star's distance. A star's distance is difficult to determine, and doing so requires a special trick which I will describe shortly. Again, an analogy might help. If you look at an aircraft flying high above the ground, it seems to move slowly, but only because it is far away. We can measure its *angular motion* across the sky, but if we don't know its distance, we can't determine its speed. It might move, for example, by one degree across the sky in a few seconds. If we knew its distance, we could convert the angular speed into a true velocity. The same is true for a star.

PEOPLE WALK AT A METER OR TWO a second, cars travel at a few tens of meters per second, and aircraft at a few hundred. Stars move through space at vastly higher speeds, anywhere from a few kilometers up to tens of kilometers a second or more, many tens of times faster than the fastest aircraft. But if they move through space so fast, why then are we unable to discern their movements by eye, even over an entire human lifetime? The reason is simply due to the fact that space is so enormous, and the distances to even the nearest stars beyond the Sun are so vast. As a result, a star's motion when viewed from the Earth is microscopic, just like an aircraft high in the sky. But while an aircraft might be moving around one thousand seconds of arc each second of time, a typical star might be moving at only a few *thousandths* of a second of arc over an entire year.

The stars, then, appear to be all but stationary with respect to each other. They seem to hold their positions because their angular motions are so microscopically small—because they are so far away. But careful observations over a very long time show that they do all move. As a result, even the constellations change their shape, albeit imperceptibly[10].

The bright stars forming Ursa Major, for example, one of the largest and most prominent of the northern constellations, known variously as the Big Dipper or the Plough, look the same now as they did hundreds of years ago— Ptolemy listed it, Shakespeare and Tennyson wrote about it, and Van Gogh painted it. And they will look just the same to our children, and to theirs. But to earliest humanity, a hundred thousand years ago, and to those equally far in the future, the constellation would be unrecognisable, grossly distorted from its present shape, the tiny motions of its seven bright stars asserting their independence over these huge stretches of time.

Except for binary stars which have an additional orbital motion, stars move pretty much in a straight line through space. This is just a consequence of Newton's first law, according to which objects remain in a state of uniform motion unless an external force is applied. Over very long periods, the star's path is bent by the collective gravity field of our Galaxy, or due to the gravitational pull of other stars, although these effects are well beyond what we can measure today.

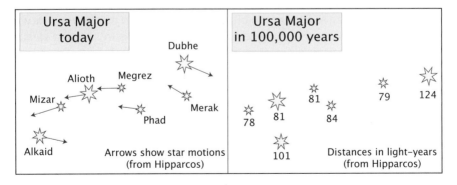

Ursa Major, now and in the far distant future

WE NEED ONE MORE important concept in our astronomical toolkit: how to measure distances to the stars. Knowledge of star distances is needed to convert angular motions into true space motions. More importantly, we need to know a star's distance so that we can convert its observed properties, such as its apparent brightness or angular radius, into true physical quantities, such as its real luminosity or linear size. It is these basic physical properties of each star which are essential in putting together a picture of its composition and its internal structure, its age and its past and future evolution.

Portrait: Julia M. Cameron

John Herschel (1867)

Here we hit a major problem, for the distances to even the nearest stars are truly vast. They are enormously, absurdly, and extravagantly large, and there are no analogies that really allow us to comprehend them. John Herschel (1792–1871), son of the illustrious William, attempted to describe the unimaginable distance scales[11]: "To drop a pea at the end of every mile of a voyage on a limitless ocean to the nearest fixed star, would require a fleet of ten thousand ships, each of six hundred tons burthen." It's an inconceivable distance, and an equally inconceivably meaningless number of peas. So let's try another analogy: the world's population of seven billion people, spaced out one every five thousand kilometers, would just about stretch to the nearest star.

This preposterous extent of space, but more poignantly its stark emptiness, are perhaps better conveyed by a scale model in which the Sun is shrunk to a marble one centimeter in size: the Earth would be a grain of salt one meter from it, and Pluto would sit far beyond at forty meters. In this grand orrery, the *nearest* stars, Proxima and Alpha Centauri, would be a staggering two hundred kilometers away.

When we come to look at the history of star position measurements we will see that astronomers struggled for centuries to measure the first stellar distances, not so surprising in view of the colossal—and unknown—problem facing them. Some of the earliest attempts to guess star distances simply assumed that they are just like the Sun, only much further away. This fairly reasonable assumption already suggested that they lay at enormous distances, and implied that accurate distance measurements would pose a most horrendous technical challenge. But the underlying assumption that all stars are just like the Sun is wildly wrong, as even the first distance measurements would reveal. While all stars are seen as if projected onto an imaginary celestial sphere, they actually lie at very different distances, some relatively nearby (although still at immense distances) and others very very much farther away. And, sadly, their brightness gives no guide as to their distance.

SO, HOW TO MEASURE the distance to a star? We certainly cannot visit them and measure their distances directly, neither can we use radar or similar techniques—they are just too far away. If we knew their intrinsic brightnesses we could infer their distances from their apparent magnitude[12], but this approach relies on a circular logic. Stars differ enormously in their innate characteristics like size, mass, and luminosity, and it is precisely these properties that astronomers want to pin down by measuring their distances. We can't deduce the distance to a town on the far horizon by measuring the angular size of a prominent church spire, without knowing the height of that particular spire. We can guess at the height from our knowledge of other church spires, and thereby hazard a guess at the town's distance, perhaps even refining it by some observed relation between height and architectural style, but this doesn't help measuring the spire's height with any degree of rigour. Distance, geographical or astronomical, is a fundamental quantity that must be measured directly.

At this point the more ardent reader might wish to reflect on how he or she would go about measuring star distances. Even after some thought the challenge might seem intractable. The solution, however, is a delightful example of scientific ingenuity, simple enough to comprehend once explained, even though its practical execution took pioneering minds and their instrumental onslaught some two hundred and fifty years to achieve. How would funding authorities today react to such a long-term proposal?

The key to measuring stellar distances is actually based on the classical surveying technique of triangulation. It simply makes use of the fact, known since the time of Copernicus, that the Earth moves around the Sun, taking one year to complete its orbit. This yearly motion provides slightly different views of space as we speed around the Sun. The nearest stars then appear to move back and forth with respect to the more distant ones over this annual cycle. The problem is that the back-and-forth motion is tiny. Miniscule, in fact. The picture of the grain of salt orbiting a marble at a distance of a meter, and using this perspective change to measure a point of light two hundred kilometers away, describes the challenge.

The effect is tiny, but is the same as that observed with our own eyes: closing first one and then the other, a nearby object appears to jump back and forth compared to a more distant object, although neither are moving. Similarly, viewed from a swiftly moving vehicle like a car or train, a nearby object like a tree appears to move rapidly against a distant object such as a church spire while, again, it is only the observer who is actually moving. The term parallax, from the Greek for change, describes the shift of angular position of two stationary points relative to each other when the observer changes position.

Parallax range finder

Our two eyes give a pair of views of the world around us, assembled into one by the brain. Although separated by only a few centimeters, this stereo vision gives us depth perception and allows us to estimate distances, at least to nearby objects. Early military range finders used optics to extend the separation between the two views to provide accurate distances for their targets much further away. Astronomers use the same stereo technique, but with views of the celestial sky separated by hundreds of millions of kilometers as the Earth moves around the Sun. In this way, Nature has generously and serendipitously granted us the possibility of measuring distances stretching across the vast expanse of our Galaxy.

In terms of the Earth's orbital motion around the Sun, each star has its own parallax angle—the ratio of the Earth–Sun distance to that of the star. If measured during this yearly motion, nearby stars appear to oscillate slightly more, back and forth, compared to the more distance stars[13]. The underlying principle of measuring stellar distances, then, is actually rather straightforward—it's just the small size of this parallax motion that makes the task so challenging.

We now know that the stars nearest to the Sun, for example Alpha and Proxima Centauri, have parallax shifts of around one second of arc, while more distant stars have smaller parallax angles still. Down the centuries, attempts to measure this parallax effect failed repeatedly because the relevant angles were so small, almost as if the stars were points of light at infinite distance. The effect was first measured, finally, only in the 1830s.

It's another of Nature's coincidences that atmospheric blurring is about the same size as the parallax shift of nearby stars. The Earth is enveloped in a shimmering shroud, which grants us a tantalising view of the Universe beyond, but which denies us any simple attempt to judge the preposterous scale on which it has been assembled.

Measuring distances to more distant stars beyond our immediate neighbours required accessing parallax angles of one tenth or one hundredth of a second of arc, and this became feasible, from the Earth, only during that latter part of the twentieth century. But measuring more distant, larger numbers, or rarer types of stars, requires measurement accuracies of around one thousandth of a second of arc. Before the advent of space measurements, this goal remained firmly beyond reach.

The parallax-based distance measurement technique is so basic that the fundamental unit of distance measurement in astronomy is based upon it. Conveying the essentials of 'parallax' and 'second' of arc it is referred to as the parsec. One parsec is simply the distance at which a star has a parallax angle of one second of arc as the Earth moves in its annual orbit around the Sun. It is the method used by parallax hunters ever since the heliocentric concept—the Earth moving around the Sun—was fully embraced some five hundred years ago. And it is the method at the heart of the Hipparcos space mission. It provides the empirical ruler of stellar distance measurements that astronomers were eager to pick up, and which they have not yet put down.

The light-year is a convenient description of distance in astronomy, and the two units are often used side-by-side (although only the parallax can be measured directly). The light-year is the distance covered by light, which travels at nearly three hundred thousand kilometers a second, over a time interval of one year. Since this terminology frequently leads to confusion, let us be quite clear that a light-year is not the measurement of time, but of distance—the distance travelled by light in one year.

Putting in the numbers, one parsec is a little more than three light years, and one light-year is some ten million million kilometers. It so happens that the nearest stars to the Sun are at a distance of a little more than one parsec, or around four light-years. So these are convenient units to describe star distances—which is why they were chosen in the first place[14].

Stellar parallax

FINALLY, LET ME GIVE SOME EXAMPLES of astronomical distances. The Sun lies about one hundred and sixty million kilometers from Earth, a trifling distance, astronomically speaking, of eight light minutes. In other words, light takes eight minutes to reach us after setting out from the Sun. Beyond our Sun, the nearest stars lie at a distance of just over one parsec, or around four light years, so that their light takes around four years to reach us.

The nearest star cluster to our Sun, the Hyades, lies at a distance of around forty parsecs, or a little more than a hundred light years. The spiral arms of our Galaxy closest to us are at around five hundred parsecs, and the centre of our Galaxy is nearly ten thousand parsecs distance, or a colossal thirty thousand light years. Our nearest neighbouring galaxies, the Magellanic Clouds, are some fifty thousand parsecs. Beyond that, great galaxy clusters stretch out to distances of tens of millions of parsecs or more—their light taking tens of millions of years to reach us.

In terms of parallax angles then, we can see the problem at hand. Astronomers needed to master measurement accuracies of around one second of arc to measure the distances to the nearest stars. And the more distant the star, the smaller the parallax. Beyond the Hyades, the next nearest star cluster to us is the Pleiades, with a parallax angle of one hundredth of a second of arc, just within the measurement horizon of Hipparcos. But to pinpoint direct distances to stars near the centre of our Galaxy would need accuracies of a few *millionths* of a second of arc, well beyond present capabilities. The next generation of space experiments, under construction, hope to conquer even this barrier within the next few years.

WE HAVE SEEN HOW METICULOUS and repeated measurement of star positions allows us to figure out their distances from Earth, as well as their velocities through space. Because of its extremely high measurement accuracy, despite the fact that its measurements were made over an interval of only three years, Hipparcos was able to chart the distances and angular motions of all the stars that it surveyed.

Now astronomers' astral telephone directory, the catalogue lists the parallax of each star, organised around its miniscule units of thousandths of a second of arc. Alongside are given their angular motions over one year, equally tiny, expressed as so many thousandths of a second of arc per year. The catalogue contains everything needed to pinpoint the stars, their distances, and their space motions. Everything, indeed, for a scientific study of what it all means.

Chapter 3

Early History

Study the past if you would divine the future.

Confucius (550–478 BCE)

M EASURING STAR POSITIONS has a long history. It's a branch of astronomy that even has its own name—astrometry, the accurate measurement of the positions of celestial bodies. Since measuring the positions of
the stars was one of the few investigations of the heavens open to the an-
cients, astronomy and astrometry were largely synonymous until a century
ago, when other types of investigation of the stars became possible.

The history of astronomy, in which astrometry plays a crucial part, is a
large and multiply-connected field. I will lay out a selective summary aiming
to pinpoint some relevant highlights as it developed from its earliest roots.
We will see that, over the centuries, improved star positions led to some remarkable and revolutionary advances in understanding our place in the Universe. One task which underpinned the field for centuries was the challenge
of determining distances to the stars. When measured for the first time in
the 1830s they revealed, at a stroke, the utter vastness of the Universe.

As in all sciences, the development of astrometry has traced out a perpetual and unwinable contest between theory and observation. New ideas
demand ever better observations to confirm them. Instrumental advances
provide new empirical evidence which in turn stimulates new ideas. Yet
along our journey, we are reminded that great advances have been punctuated by long periods of scientific hibernation or decline. History provides
a valuable backdrop, if only lest we are tempted to think that what is surely
today a golden age can never suffer comparable setbacks.

M. Perryman, *The Making of History's Greatest Star Map*, Astronomers' Universe
DOI 10.1007/978-3-642-11602-5_4, © Springer-Verlag Berlin Heidelberg 2010

THE FLOURISHING OF WESTERN ASTRONOMY over the past few hundred years has its origins in much earlier bursts of scientific activity[15]. Prehistoric sites revealing celestial alignments, such as Newgrange in Ireland and Stonehenge in England, date from around 3000 BCE.

The first recorded developments emerged in Mesopotamia around 1000 BCE where, in the land between the Tigris and Euphrates rivers now occupied by southern Iraq, Assyro–Babylonian astronomers systematically observed the night skies, building on common lore already conscious of the changing daylight over the year. They observed, they measured, and they recognised, for the first time, that certain celestial phenomena were periodic: amongst them the regular appearance of Venus, and the eighteen year cycle of lunar eclipses. Their careful records formed the basis for later developments, not only in ancient Greek and Hellenistic astronomy, but also in classical Indian and in medieval Islamic astronomy.

Plato

Early Greek philosophers, the Pythagoreans amongst them, played a key part in astronomy's earliest awakening. They believed that the underlying regularities, or laws of nature, were discoverable by reason. As part of this philosophical school, astronomers of ancient Greece tried to understand the Universe based on principles of 'cosmos', or order. The revelationary idea that the Earth might be spherical began to replace the pre-Socratic view that its surface was flat.

Plato (427–347 BCE) and his contemporaries knew that the heavens rotated night after night with constant speed, the 'fixed' stars preserving their relative positions as the heavens turned. Moving amongst them in a complex and unfathomable way were the seven wanderers—the Greek *planetes*—the Sun, the Moon, and the planets visible to the naked eye: Mercury, Venus, Mars, Jupiter and Saturn. Seen from Earth, their positions trace out complex and convoluted paths, sometimes with even backward 'retrograde' loops. As described by Goodman & Russell[16]: "Their erratic behaviour had baffled and infuriated generations of Greek thinkers, up to Plato himself. It seemed impossible to reconcile their celestial meanderings with either the supposed divinity of heavenly bodies or with any simple concept of circular motion."

Scientific thinking was dominated by the idea that the Earth lay fixed at the centre of the Universe. This fundamental tenet in mankind's early views completely obstructed the correct interpretation of planetary motions. We now know that the apparently complex paths of the planets follows from the rotation of the Earth, combined with the orbits of the Earth and other planets around the Sun. When interpreted correctly in a heliocentric system, and with elliptical rather than circular orbits, the motions are simple. But in a system in which the Earth is fixed they are not. Heraclides had hinted at a Sun-centred system in the fourth century BCE, but his view failed to find support in a culture generally attached to the idea of an Earth fixed in space, which would continue to hold sway, erroneously, for a further two millennia. Perhaps we should really not be too surprised by the reluctance of the ancient Greeks to accept our present very non-intuitive idea of a rapidly spinning Earth careering through space.

Aristarchus of Samos (circa 310–230 BCE) made one of the first attempts to determine the distances and sizes of the Sun and Moon. He deduced the ratio of their distances using trigonometry, by measuring the angle between them when the Moon is exactly half lit. He also argued in favour of the heliocentric, Sun-centred system, a view supported by Seleucus of Seleucia around the second century BCE. But these ideas found little favour at the time, and they remained lost amongst the geocentric, Earth-centred system still being championed by most of his contemporaries.

To explain the complex apparent motions of the planets and the varying speed of the Moon, geocentric proponents could not appeal to planetary orbits which were simply circular. They had to introduce complex epicyclic motions—patterns traced out by circles turning around the circumference of larger circles. Contrived though they were, they broadly explained the irregular speeds of the planets across the sky throughout the year, occasionally even tracing backward loops with respect to the background stars.

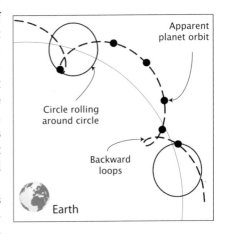

Planet movements erroneously explained by epicycle motions

At around the same time, Eratosthenes (276 BCE–194 BCE) invented a system of latitude and longitude, and used the varying elevation of the Sun to estimate the size of the Earth, deriving a value which would be used for centuries afterwards.

Hipparchus (active circa 160–126 BCE) is credited with a number of important advances in astronomy, although most of what is known about his work is handed down from Ptolemy's second century Almagest (the name itself assigned by ninth century Arabic translators). He pioneered the classification of star brightnesses still in use today, dividing them into six groups, the brightest designated as first magnitude (the first to be seen at dusk), and the faintest as sixth. He followed the ancient Babylonians in dividing a circle into three hundred and sixty degrees, each of sixty minutes of arc, and he compiled the first systematic star catalogue, recording star positions with an accuracy of about one degree. He was the first to describe the precessional motion of the fixed stars, that is the steady wobbling of their positions over decades due to the steady change in the position of the Earth's spin axis in space, just like a spinning top.

Library at Alexandria

Wikimedia Commons

But Hipparchus incorrectly continued to uphold the geocentric system. His argument was that a precisely circular orbit of the Earth around the Sun failed to explain the planetary motions. We know now that the planetary orbits are elliptical, so that his argument was compelling, but fallacious. Nevertheless his views, and his considerable authority, effectively ensured that the heliocentric hypothesis would lay discarded for many centuries. More than two hundred years later, in the second century CE, Ptolemy would put forward his own variant of this geocentric world view, and would also invoke the complex epicyclic motions to predict, successfully even if based on flawed models, the future positions of the planets.

Greek scientific activity came to a reasonably abrupt end, for reasons which society in the twenty first century might like to take note. According to Goodman & Russell: "...the most likely reasons seem to be the paucity of scientists and their isolation. The Athenian institutions and the Alexandrian Museum were rarities. Over the Greek centuries most scientific activity was uncoordinated, and poor communications often resulted in duplication of effort or ignorance of what had been achieved. Education in Greek schools concentrated on music, poetry and gymnastics, not on science; with the exception of Alexandria there was no strong government or social encouragement of the sciences. For Europe to have developed the sciences further from these Greek foundations, knowledge of Greek, close contact with the Greek scientific texts, and sustained interest in what they might teach were all necessary. But in the centuries after the fall of the Roman Empire in the west, none of these conditions was satisfied."

The subsequent decline of the Roman empire—in population, economic and political order—precipitated by barbarian attacks, decimating epidemics, and inability to provide for the succession of government—ushered in the 'Dark Ages'. Europe disappeared under the likes of the Vandals and Visigoths for centuries.

TO THE EAST, MEANWHILE, China's first major economic burst under the Han dynasty (206 BCE–220 CE) nurtured a philosophical period roughly coincident with the innovative centuries of Greek philosophical and scientific thought. In India the Gupta empire, from around 320 CE, also stimulated navigation and advances in numeracy. Astronomy was recognised as a separate discipline, and around 500 CE Aryabhata held that the Earth was a sphere rotating on its axis. Much later in China, under the Sung emperors in the eleventh and twelfth centuries, a distinct advance in scientific achievement was accompanied by the flourishing of observational astronomy. Patronage from the emperors ensured the astrological fortunes of their dynasty, as well as regulation of the official calendar. Star catalogues, as well as records of sunspots and comets, have been handed down to us from this time. But China's separation from the west, buffered by the nomadic tribes of central Asia, meant that their developments had little immediate or direct influence on Europe's subsequent scientific re-awakening.

The burgeoning Islamic culture was to dominate the world's economic development from the early part of the seventh century for the next three hundred years. Geographically closer than China and India, it had more of a direct influence on the west, and played an important part in reviving scientific enquiry in Europe. Supported by the patronage of the Caliphs, Islamic scholars transmitted, translated, and criticised anew the ancient Greek scientific texts. Knowledge of astronomy was also inspired by practical needs: to establish each mosque's direction to Mecca, the timing of daily prayers, and to rule on the precise beginning and end of Ramadan.

Ulugh Beg's observatory, Samarkand

Amongst notable achievements in astronomy, Al Battani, working around 900 CE at his observatory on the Euphrates, refined Ptolemy's description of the orbits of the Sun and Moon. Ibn Yunus (circa 950–1009) described planetary alignments and lunar eclipses accurate enough for a great figure in late nineteenth century astrometry, Simon Newcomb, to use them for his own theories of lunar motion. Ulugh Beg, grandson of the Mongol conqueror Tamerlane, constructed a sextant of thirty-six meter radius in Samarkand in 1428, sited in a circular arc between marble walls. He used it to compile a catalogue of 994 stars. With star positions accurate to about one degree, it is generally considered as the greatest star catalogue between those of Hipparchus and Tycho Brahe.

EUROPE'S SLOW EMERGENCE from the 'Dark Ages' began to gather pace under the Carolingian dynasty, from around 730 CE onwards. Trade and towns in western Europe started to revive, and economic life progressively shifted from the Mediterranean to the North Sea and Atlantic coast. From this renewed prosperity, and improved political stability, the foundations of the modern age slowly emerged. Indeed a significant body of scholarship over the past fifty years has quite dispelled the idea that Europe between 500–1500 CE was intellectually and technologically moribund.

Nicholas Copernicus

A new curiosity about the heavens surfaced and thrived. The basic imponderable of astronomy and cosmology until the Middle Ages, that of the inexplicable motion of the five planets known at the time, was picked up again after a pause of a full millennium. Nicholas Copernicus (1473–1543) openly drew on this rich medieval tradition, and finally laid the secure foundations for a credible heliocentric world model, in which the Earth moves in orbit around the Sun rather than vice versa. Little new observational evidence motivated his thinking: he lived before the invention of the telescope, and his best observational accuracy was only about ten minutes of arc. Rather, rediscovery and reinterpretation of the ancient texts played a major part in the origins of Renaissance culture in general, and astronomy in particular. According to the account of Thomas Heath[17]: "Copernicus himself admitted that the [heliocentric] theory was attributed to Aristarchus, though this does not seem to be generally known."

Far from being fixed in space, Copernicus proposed that the Earth was subject to three kinds of motion. The first was an annual orbit around the Sun. The second was a daily rotation accounting for day and night, but about an axis tilted with respect to its orbit plane which would account for the changing character of the seasons. Third was a more complex and very long period wobble of the Earth's axis as it spins, known as precession. His *De Revolutionibus Orbium Coelestium*, Concerning the Revolutions of the Heavenly Bodies, arguably marks the beginning of Europe's scientific awakening.

The acceptance of a Sun-centred solar system accounted for the most extreme contributions to the backward looping motions of the outer planets. But Copernicus still needed the highly contrived epicycles to explain the detailed planetary motions, albeit of a smaller magnitude than those which had to be invoked by Ptolemy in his Earth-based system. Yet other subtleties were needed to match the known orbits of the planets, for Copernicus was erroneously trying to fit a series of circular motions to their yet-to-be discovered elliptical paths.

Only with the later work of Johannes Kepler, Galileo Galilei and Isaac Newton, and the realisation and understanding that the planetary orbits were elliptical rather than circular, could the need for epicyclic motions be discarded altogether. By demonstrating that the motions of celestial objects could be explained without putting the Earth at rest in the centre of the Universe, the work of Copernicus stimulated further scientific investigations and became a landmark in the history of modern science.

In 1610, Galileo Galilei published his *Sidereus Nuncius*, which described the surprising observations made with his newly-invented telescope—mountains on the Moon, moons around Jupiter, and patchy nebulae for the first time resolved into innumerable faint stars. His support for Copernican heliocentrism set in train a lengthy conflict with the Catholic Church, leading to suggestions of heresy, perhaps intellectual rather than theological, and his eventual trial and house arrest in 1633.

Galileo before the Holy Office

Giordano Bruno (1548–1600) was a proponent of heliocentrism and the infinity of the universe, earlier burned at the stake albeit for other more extreme theological heresies. Such were the harsh penalties for questioning the authority of the Holy Scriptures, which decreed that the Earth was the centre of the Universe, and that all heavenly bodies revolved around it. From the altar of St Peter's Basilica in Rome in March 2000, Pope John Paul II issued an apology for the errors of the Church over the last two millennia, including the trial of Galileo: "The error of the theologians of the time, when they maintained the centrality of the Earth, was to think that our understanding of the physical world's structure was, in some way, imposed by the literal sense of Sacred Scripture."

Johannes Kepler (1571–1630) appeared as an important figure in the seventeenth century astronomical revolution, best known for his eponymous laws of planetary motion. He defended heliocentrism from both a theoretical and theological perspective. His observational work with Tycho Brahe encouraged his own repeated attempts to calculate the full orbit of Mars around the Sun. Eventually, in 1605, he found that while a circular orbit did not match the observations, an elliptical one did. It was a simple answer which he had previously assumed to be too straightforward for earlier astronomers to have overlooked. He concluded that all planets move in ellipses, with the Sun at one focus. This deduction, his first law of planetary motion, provided a foundation for Isaac Newton's theory of gravitation.

Aside from his mathematical skills, Kepler lived at a time when there was no clear separation between the science of astronomy and the pseudoscience of astrology, and he also had a reputation as a skilful astrologer.

Portrait: Godfrey Kneller

Isaac Newton

Isaac Newton (1642–1727) occupies a lofty pedestal unrivalled in the history of science, and his *Philosophiae Naturalis Principia Mathematica* of 1687 is arguably its most influential book. He bestowed on mathematics and physics a rich and complex collection of ideas. Together, and in a stroke, his laws of motion, gravitational attraction, and the inverse square law of gravity gave an explanation of the motions of all celestial bodies. But this package of new ideas, Newtonianism, was not the only scientific movement competing for support, and it was not accepted immediately. Rivals included Hutchinsonianism in England, centred around the Trinitarian theology of John Hutchinson, and Cartesianism in France, based on the influential philosophical doctrine of René Descartes.

THE HELIOCENTRIC HYPOTHESIS eventually prevailed, and Newtonian gravity along with it. With their joint acceptance came an inevitable consequence, a conclusion that would mark a fundamental turning point in science. For if the Earth indeed moves in orbit around the Sun, then the 'fixed' stars cannot remain truly fixed in space. Unless they were at infinite distance, they would have to possess a parallax motion—an oscillation of their *apparent* position which would arise from the Earth's annual motion around the Sun. To be sure, neither Aristarchus nor Copernicus had observed the effect, and this fact alone implied that the distances to the stars must dwarf even the colossal distance scale of the solar system.

The conclusion that the parallax effect had to exist now seemed inescapable. A renewed push to detect it began, armed with the certain knowledge that the effect being sought would be tiny. Great improvements in measurement accuracy would be needed before the effect could be measured. The warm-up was over, and the race to measure parallax began in earnest.

STARTING SOME THREE or four centuries ago, the search for parallax, the further comprehension and definitive acceptance of Newtonianism, and understanding the precise nature of the Earth's motion through space were interwoven, and together motivated the progressive improvement of angular measurements. A related but more urgent practical problem came to a head at the same time: the navigational problems associated with the determination of longitude.

For most of history, explorers in general and mariners in particular had struggled to determine their precise longitude, their point east or west of some reference point on the Earth. Latitude has the Earth's equator as a natural reference plane, and it can be determined by observing the altitude of the Sun or stars using specialised protractor-like instruments like the quadrant or sextant, or the astrolabe, a sort of analogue calculator capable of working out different kinds of problems in spherical astronomy. There is, however, no such unique reference position for longitude, and no practical means for its direct estimation. For a ship lost at sea on the slowly-spinning Earth, estimating longitude was frequently a matter of life or death. But it was tied directly to the knowledge of time. Without time, there was no hope of determining longitude.

The problem was urgent, and the economic consequence of ships, cargos and lives lost at sea was substantial. In France, Louis XIV promoted the construction of the Paris Observatory, established in 1667 under director Giovanni Domenico Cassini, with the express purpose of extending France's maritime power and expanding her international trade. In England, King Charles II was similarly moved to found the Royal Greenwich Observatory in 1675, with the purpose of compiling detailed star maps for navigational purposes. He instructed the first Astronomer Royal, John Flamsteed, "to apply himself with the most exact care and diligence to the rectifying of the tables of the motions of the heavens, and the places of the fixed stars, so as to find out the so much-desired longitude of places for perfecting the art of navigation." In 1725 Flamsteed's *Historia Coelestis Britannica* was published posthumously, and contained his catalogue of 2935 stars. It was the first significant contribution of the Greenwich Observatory, and a landmark in the history of stellar astrometry—positions, accurate to around ten or twenty seconds of arc, were the first measured with telescopic sights, and a major improvement over earlier work.

Star charts alone, however, could not provide a solution to the problem of navigation. Without a clock that could keep accurate time over months of an ocean voyage, there was no practical way of establishing what time it was at the reference point. With Galileo's discovery of the four brightest moons of Jupiter in 1610, named by him as the Medicean stars after his patron but subsequently named the Galilean moons in his honour, it became possible in theory to deduce the time on board ship by observing when the satellites appeared from behind the planet—the events occurred frequently and, more importantly, predictably. The world's first national almanac, the *Connaissance des Temps*, giving these eclipse timings, was published from 1679. Tables could then be consulted to see when these events were due to occur as measured at the prime meridian. Galileo himself pursued this approach to navigation during his lifetime, and even petitioned King Philip III of Spain who had also offered a financial reward for a breakthrough in determining longitude. Yet such measurements could only be made at night, were much at the mercy of the weather, and quite impossible from a rolling boat in high seas, and it failed to provide a practical solution. Before the middle of the eighteenth century, most sailors continued to use a variant of dead reckoning to try to keep track of their position. Galileo died in 1642, before his method became widely used by cartographers on land.

John Harrison's chronometer H5

The search for a solution was spurred on by the Longitude Act of 1714. The British Parliament offered a prize of £20 000, a fortune of some £6 million in present worth, for a method that could determine longitude within thirty nautical miles. A solution was eventually found through the use of accurate celestial charts and lunar tables, in combination with the measurement of precise time.

With the success of the marine chronometer in the 1760s, pioneered by English clockmaker John Harrison, time could at last be carefully measured and accurately transported throughout a long voyage. Accurate clocks eventually became commonplace. The problem of navigation at sea was considered as solved, and the Board of Longitude was dissolved in 1828[18]. Not until 1884, however, was the International Meridian Conference meeting in Washington DC to adopt the meridian passing through Greenwich as the universal, if quite arbitrary and long contested, zero point of longitude. France abstained, maintaining her preferred use of the Paris meridian until 1911 for timekeeping purposes, and until 1914 for navigation.

From 1767, the Nautical Almanac has been published annually. From 1958, the US Naval Observatory and the HM Nautical Almanac Office have jointly published a unified volume, for use by the navies of both countries. It still tabulates the positions of the Sun, moon, planets, and a number of stars selected for ease of identification and widely spaced across the sky. To find the position of a ship or aircraft by celestial navigation follows the method unchanged for more than two centuries: the navigator uses a sextant to measure the height of a chosen object above the horizon, notes the time from a chronometer, and deduces location by comparing the star's position with that given in the almanac for that particular time.

Thousands of lives and considerable fortunes had been lost before star charts in combination with transportable time could be used for reliable navigation. Perhaps astronomical research might have been financed better down the centuries if just a fractional royalty of the commerce subsequently piloted to safety had accrued to its own coffers.

Well into the 1800s, star positions provided the most accurate means of determining geographical coordinates, and with them the distance between cities or the position of national borders. An interesting parallel occurs today: the huge civilian, commercial, and military reliance on global satellite navigation, notably GPS, depends crucially upon the inclusion of the delicate effects of Einstein's special and general relativity: omit them from consideration, and positions would be

Sextant of Johannes Hevelius

several kilometers in error after only a few hours. In this area alone, astronomy and relativity have proven indispensable to this important social and commercial venture. The cost of Europe's own global satellite navigation system, also named Galileo, and now nearing implementation, wouldn't quite pay the salaries of the world's astronomers for the last hundred years, but the numbers aren't too far apart.

Unknown spin-offs that might emerge, perhaps way down the line, is an argument often used for societies to support pure research.

As COPERNICANISM SPREAD THROUGHOUT EUROPE, and the heliocentric cosmos gained acceptance, the race to measure parallax gathered pace. Even before 1600, astronomers were in agreement that the crucial evidence needed to detect the Earth's motion around the Sun was the measurement of trigonometric parallax. The early British Astronomers Royal, for example, appreciated the importance of measuring stellar distances, and had devoted much energy and ingenuity to the task. But it was to take a further two hundred and fifty years, and failure upon failure, until the first star distances were measured.

From ancient times through to the start of the twentieth century, the measuring of celestial positions had always been central to astronomical research. The quality of the instruments determined the accuracy of the measurements. The art of dividing a physical circular scale into degrees and minutes of arc was but a practical problem, essentially one of accurately marking off successively smaller angles. But it was one of such technical complexity that it now presented the principal barrier to advancing research[19].

Statue of Tycho, Hven

Michael Perryman

During the Middle Ages, European and Islamic astronomers adopted a brute force approach to the problem. They constructed observing circles with very large radii such that they could more easily inscribe and further dissect more precise angles on their annular limbs. Tycho Brahe (1546–1601), whose observations provided the basis of Kepler's laws of planetary motion, employed such an instrument. His Great Quadrant had a radius of fourteen cubits, around seven meters, and probably reached an accuracy of around six minutes of arc, one fifth the Moon's diameter. At his lavish observatory of Uraniborg on the Danish island of Hven, developed under the patronage of Frederick II, King of Denmark and Norway, he used his families of sextants, armillary spheres, and quadrants. By the last decade of the sixteenth century, he was reaching an unsurpassed accuracy of around twenty seconds of arc.

Despite his observational skills and his extravagant funding, Tycho attempted, but also failed, to detect parallax motion. But the accuracies that he achieved allowed him to deduce that the stars must lie several thousand times more distant than the Earth from the Sun. These distances were so frighteningly immense that he was convinced Copernicus must be in error, and that the Earth was indeed fixed at the centre of a modified 'Tychonic' system. In reality, with even the nearest stars having a parallax angle of only one second of arc, Tycho's accuracy was still twenty times too poor, and even his careful measurements could not but have failed to detect its effects. Nevertheless by the end of the sixteenth century, his catalogue of a thousand stars, and a similar effort by Landgrave (Baron) Wilhelm the Wise of Hesse (1532–1592), set the standard for future surveys.

The sextant and quadrant were protractor-like instruments designed to measure angles between pairs of stars, of up to sixty and ninety degrees respectively. Catalogues were built up from many pairs of separations. Portable versions were later fixed in the meridian plane—the imaginary circle perpendicular to the celestial equator and horizon. Observations with wall-mounted 'mural' instruments began with Tycho's large meridian quadrant. Fixed to the local horizon, stars appear to drift past the local meridian as the Earth spins: this gave one part of the star's coordinates (the equivalent of geographical longitude, or right ascension) from the timing of its transit, and the other (the geographical latitude, or declination) from the graduated instrument itself. These were later replaced by the meridian circles, consisting of a horizontal axis in the east–west direction resting on fixed supports, about which a telescope mounted at right angles could revolve freely.

Until the late eighteenth century, the art of graduating circular scales into ever finer subdivisions was pursued in earnest, but carried out largely in secrecy to thwart the competition. Wider exposition of practical methods accelerated when the Board of Longitude, which had been formed in 1714 to solve the problem of finding longitude at sea, persuaded John Bird to publish his methods in 1767. In the following decades, Jesse Ramsden and Edward Troughton continued the advance of angular measurements. Prestigious Fellowships of the Royal Society were awarded for their instrument advances, underlining the importance with which the measurement of stellar positions was held, and testament to their innovation. In

Bradley's quadrant from 1750

his chronicle of the rise and fall of economies throughout history, Peter Jay includes Ramsden's dividing machine for accurate graduation of circles for navigational and surveying instruments as one of the inventions which contributed to the productivity gain that signaled the Industrial Revolution[20].

During the later parts of the seventeenth and early eighteenth century, other instruments were added to the arsenal of techniques for measuring star positions. These included the transit telescope, which added a regulator clock to time the passage of stars across the Earth's meridian. Its more specialised form, the zenith sector, was used by Robert Hooke (1635–1703), one of the most important scientists of his age, in his own attempts to measure the parallax of the bright star Gamma Draconis.

Gamma Draconis is a giant star in the constellation of Draco, and a notable object throughout recorded history. According to Allen[21]: "Its rising was visible about 3500 BCE through the central passages of the temples of Hathor at Denderah and of Mut at Thebes. And Lockyer [Sir Joseph Norman Lockyer, 1836–1920] says that thirteen centuries later it became the orientation point of the great Karnak temples of Rameses and Khons at Thebes, the passage in the former, through which the star was observed, being 1500 feet in length; and that at least seven different temples were oriented toward it. When precession had put an end to this use of these temples, others are thought to have been built with the same purpose in view; so that there are now found three different sets of structures close together, and so oriented that the dates of all, hitherto not certainly known, may be determinable by this knowledge of the purpose for which they were designed. Such being the case, Lockyer concludes that Hipparchus was not the discoverer of the precession of the equinoxes, as is generally supposed, but merely the publisher of that discovery made by the Egyptians."

Portrait: Thomas Murray

Edmond Halley

The interest of the star Gamma Draconis to the seventeenth and eighteenth century parallax hunters was simply that it lay almost exactly in the zenith of Greenwich, minimising refraction by the atmosphere, and conveniently studied by a fixed telescope pointing straight up—Hooke had cut a hole in the roof of his apartment to observe it. In 1674 he claimed the detection of a parallax for Gamma Draconis of roughly thirty seconds of arc, and with it proof of the Copernican system, although later work showed that his results were in error.

A remarkable and crucial breakthrough came in 1718. Edmond Halley, who had been comparing contemporary observations with those that the Greek Hipparchus and others had made, announced that the bright stars Aldebaran, Arcturus, and Sirius were displaced from their expected positions by large fractions of a degree. He deduced that each star had its own distinct velocity across the line of sight, or proper motion. It was the first convincing experimental suggestion that stars were moving through space.

Halley's scientific achievements were many and varied. He predicted the return in 1758 of a periodic comet which now bears his name, identified solar heating as a cause of atmospheric turbulence, and suggested a measurement of the distance between the Earth and the Sun by timing the transit of Venus. Less successful was his suggestion, to explain anomalous compass readings, that the Earth was a hollow shell some eight hundred kilometers thick. Pioneers in science do get it wrong. This example also shows the limits in scientific understanding that existed a mere three hundred years ago.

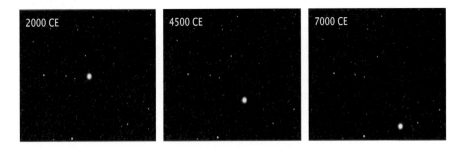

*The motion of Arcturus, measured by Hipparcos, followed from its
present position (left), to its predicted position 5000 years from now (right)*

BY 1725, INSTRUMENTAL ADVANCES had reached an accuracy of a few seconds of arc. The Reverend James Bradley, England's third Astronomer Royal, was deeply immersed in his own efforts to measure parallax, and was also focusing his attention on Gamma Draconis. His attempts were unsuccessful, for the star is too distant for the effect to show up at the accuracy then available. But they pushed his own estimates of the nearest stellar distances out to nearly half a million times that of the Earth from the Sun.

More importantly, Bradley's experiments yielded an unexpected surprise: the detection of a small systematic shift in his star positions, of a form very different from that expected from the effects of parallax, and which he eventually correctly attributed as resulting from the addition of the velocity of light to the Earth's velocity as it moves in orbit around the Sun. The usual analogy is that when rain is falling straight down, and you're walking briskly ahead, you tilt an umbrella forward slightly to intercept the apparent direction of the rainfall. It's a consequence of adding two velocities. Dinghy sailors know the effect well: the flag atop the mast doesn't indicate the wind direction, but that of the wind and boat speed combined. Bradley had pondered the meaning of his perplexing star measurements for three years before enlightenment struck, his insight precipitated by observing such a moving vane on a sail boat on the River Thames.

Bradley's observations of this effect, known as stellar aberration, or the aberration of starlight, was announced in 1729, and arguably rates as one of the most significant discoveries in the history of astronomy. It provided the first direct proof that the Earth was moving through space. His results therefore supported the Copernican theory, that the Sun, rather than the Earth, was the centre of the solar system. But it confirmed, at the same time, Danish astronomer Ole Rømer's discovery of the finite velocity of light fifty years earlier. Rømer had been observing the eclipses of Jupiter's moons as part of the ongoing challenge to establish a practical method to determine longitude. His

James Bradley

own conclusion that the velocity of light was finite, rather than propagating at infinite speed, wasn't fully accepted until Bradley's measurement of aberration provided crucial supporting evidence.

By failing to detect the parallax of Gamma Draconis, even at the unprecedented level of about one second of arc, Bradley's observations went further in confirming Newton's hypothesis of the enormity of stellar distances, and confirmed that the measurement of parallax would continue to pose a technical challenge of inordinate delicacy.

Nevil Maskelyne, England's fifth Astronomer Royal, spent seven months on the remote volcanic island of Saint Helena in 1761, a crucial staging and rendezvous point for sailing ships in the South Atlantic. He had been despatched by the Royal Society to observe the transit of Venus, and thereby to improve knowledge of the Earth's distance from the Sun and the scale of the solar system. He used a zenith sector and plumb-line in an unsuccessful attempt to measure the parallax of Sirius during the same expedition.

During the eighteenth century, after Halley's first detection of stellar motions, the movements of many more stars were being announced. In 1783 William Herschel found that he could partly explain these collective motions by assuming that, in addition to the Earth's motion around the Sun, the Sun itself was moving through space. With his sister Caroline, Herschel made numerous important advances: he discovered Uranus in 1781, two moons of Uranus and two of Saturn between 1787–89, and discovered infrared radiation. He observed and catalogued binary stars, detecting the first orbital motions and, in the process, the first proof that Newton's laws of gravitation applied outside the solar system. He was a prolific telescope maker, and also sought to detect a parallax shift from measurements repeated over the course of a year. Yet in this, even armed with his largest telescope, a primary mirror more than a meter in diameter and a colossal twelve meter focal length, he too failed. As he wrote in 1782: "To find the distance of the fixed stars has been a problem which many eminent astronomers have attempted to solve; but about which, after all, we remain in a great measure still in the dark."

Meanwhile, another important step in expanding ever larger star surveys was the work of Jérôme Lalande (1732 –1807) in France. His *Histoire Céleste Française* of 1801, gave the places of 50 000 stars with an accuracy of around three seconds of arc[22].

The symbolic if arbitrary figure of one second of arc was now within sight, and attempts to measure parallax intensified. But since the distances to even the nearest stars were still unknown, nobody could predict what angular accuracy would be needed for the effect to be detected. The topic was the focus of many learned papers published in the opening decades of the 1800s. The failures of Tycho, Hooke, Flamsteed, Bradley, Maskelyne, Herschel and many others, were followed by a renewed flurry of measurements and false claims: amongst them by Giuseppe Piazzi in Palermo, Giuseppe Calandrelli in Rome, François Arago in Paris (later Prime Minister of France), Baron Bernhard von Lindenau in Gotha, Johan Schröter in Lilienthal, and John Brinkley in Dublin. In the words of Alan Hirshfeld[23]: "Each claimed victory in what astronomers increasingly perceived as a parallax race. But instead of glory, the recent parallax competitors gained only the suspicion, if not the contempt, of their colleagues."

What was urgently needed were criteria for selecting stars likely to be close to the Sun, to avoid time wasted in trying to measure distant stars. In 1837, German-born Wilhelm Struve, working at Dorpat in Russia (now Tartu in Estonia), gave three suggestions: the star should be bright; it should be moving with a large angular rate across the sky (although this *could* be a rapidly moving star at a large distance, it was more likely to be 'nearby'); and if the star was one of a binary pair, the two components should be well separated as judged by the time taken to orbit each other. Struve drew up a list of stars satisfying these criteria. Our present-day knowledge confirms that astronomers were, at last, able to select some of the very nearest stars on which to focus their painstaking measurements.

Wilhelm Struve

AFTER MANY UNSUCCESSFUL ATTEMPTS, the very first stellar parallaxes were measured and reported during a burst of activity in the 1830s, two hundred years after Isaac Newton had removed any final doubt that the Earth was in motion around the Sun. After this protracted marathon to detect the first parallax, three scientists breasted the winning tape almost together.

Wilhelm Struve had selected the bright, high proper motion star Vega for study. At his disposal in Dorpat was a twenty-four centimeter aperture refractor, manufactured by the German physicist and craftsman Joseph Fraunhofer, and the largest instrument of its kind in the world. Equipped with a 'filar micrometer', long used for measuring separations of double stars, two tiny parallel wires or threads, often of fine but immensely strong spider silk[24], could be moved by the observer using a screw mechanism. The slowly changing separations between the target star and a nearby comparison star could be tracked.

Struve's refractor, Tartu

Struve's results from seventeen observations starting in 1835 were announced two years later, giving a parallax of one-eighth of a second of arc, close to the present value. But since there had been a long history of fallacious claims to the measurement of parallax, others remained sceptical, and Struve continued his measurements until, in 1840, he gave the results from nearly a hundred observations.

Friedrich Bessel is generally credited as being the first to publish a reliable parallax, spurred on in his measurements by correspondence with Struve and the latter's preliminary result for Vega. From observations made between 1837–38, Bessel carefully tracked the detailed path of the fast-moving binary star known as 61 Cygni, using the heliometer at Königsberg (now Kaliningrad), also manufactured by Fraunhofer. The heliometer had originally been designed to measure the Sun's angular diameter, and hence the name. Its sixteen centimeter diameter refractor lens had been sliced in half, each segment mounted side-by-side, so forming a pair of images which could be adjusted laterally by turning a thumbscrew. Bessel used it to follow the slowly changing angles between his chosen target and a comparison star close by on the sky. Careful monitoring over the course of a year would show a varying separation if the accuracies were sufficient to discern the parallax wobble of the nearby binary.

Wikimedia Commons

Friedrich Bessel

In Alan Hirshfeld's delightful account of this protracted race, he describes Bessel's precision instrument as "almost painfully beautiful: a copper-shaded, mahogany-veneer tube; burnished knobs, gears, and wheels; and a wooden equatorial mount that descended to Earth through a complex of gracefully splayed struts and stout beams." To guarantee stability "the central part of the [telescope] tower's base was filled with five feet of masonry. Atop this were slabs of sandstone and a layer of timbers. Bolted to the timbers were a series of iron-reinforced beams that rose to the upper reaches of the tower and supported the platform on which the heliometer rested."

It was an excellent piece of engineering, and with it pointed to the heavens the first star parallax was measured: in 1838, Bessel announced that 61 Cygni had a parallax of 0.314 seconds of arc, placing it at a distance of three parsecs, or ten light-years. What convinced others that a star distance had been measured for the first time was the match between theory and the expected pattern of separations as the Earth moved in its annual orbit around the Sun.

Hot on Bessel's heels was the work of Thomas Henderson, first Astronomer Royal for Scotland, who published a parallax for the nearby star Alpha Centauri in 1839, derived from observations made even earlier in 1832–33 at the Cape of Good Hope. Although the star is particularly close to the Sun, and its parallax angle therefore amongst the very largest of all stars in the sky, it is only observable from southern latitudes. And with the exception of occasional southern expeditions, such as Halley's and Maskelyne's to

Saint Helena, and Abbé Nicolas Louis de Lacaille's catalogue of more than ten thousand stars observed from the Cape of Good Hope in the 1750s, the southern skies had received but scant attention. The situation was addressed by England's Board of Longitude which set up a dedicated observatory at the Cape under its first director, the Reverend Fearon Fallows, whom Henderson replaced in 1832.

Henderson returned to England barely a year later, dissatisfied with working conditions at the Cape. But included amongst his observations, made with an ordinary mural circle and yet to be analysed, were a series of careful measurements of Alpha Centauri. The star was bright, with a large proper motion, and also one component of a binary with a large separation. It thereby handsomely fulfilled all three of Wilhelm Struve's criteria of likely proximity.

The announcements of Bessel and Struve, and the star's probable proximity, prompted him to re-examine his own observations from which he duly determined its parallax. Still today, the binary pair of Alpha Centauri, and their fainter companion Proxima Centauri, remain the nearest known stars to our Sun. Pin-pointed from the Hipparcos space measurements, Alpha Centauri has a parallax of 0.742 seconds of arc, which corresponds to a distance of 1.35 parsecs, or 4.396 light-years—just over forty million million kilometers.

What had at last come together was the understanding that star distances could be measured using the Earth's motion around the Sun, and the pin-pointing of those most promising to measure on account of their likely proximity. Improvements in telescope size, quality, and accuracy, inspired and drove the relentless pursuit.

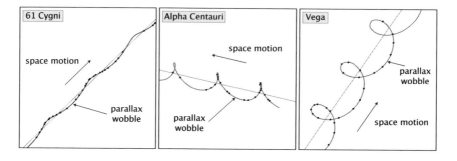

Motions of 61 Cygni, Alpha Centauri, and Vega over three years from Hipparcos
(small dots show the positions measured from space at different times)

THESE FIRST PARALLAX MEASUREMENTS provided the very first rigorous determination of the distances to the stars. The confirmation that they lay at very great, yet not infinite, distances represented a turning point in the understanding of the Universe. The moment when distances to the stars, and the enormous scale of space, were suddenly and unambiguously revealed must rank as one of the most pivotal in the entire history of science.

John Herschel, President of the Royal Astronomical Society at the time, congratulated members of the society that they had[25] "lived to see the day when the sounding line in the universe of stars had at last touched bottom." In awarding the society's gold medal to Friedrich Bessel in 1841, he described it as "the greatest and most glorious triumph which practical astronomy has ever witnessed."

The refractor used by Struve to measure the parallax of Vega still resides in the museum of the Old Observatory in Tartu, Estonia. Bessel's heliometer, along with the observatory and city of Königsberg, was destroyed in the wartime ravages of 1944–45.

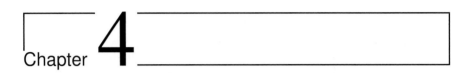

Chapter 4

Developments 1850–1980

Such cooperation proves, if proof is necessary, that science knows
no nationality, and that common pursuit of truth for truth's sake
affords one touch of nature which makes the whole world kin.

Sir David Gill (1843–1914)

O VER THE PERIOD OF THREE HUNDRED years leading up to the detection of the first parallax in 1838, the measurement of star positions had actually followed two somewhat separate branches. The first of these concentrated on the measurement of parallax, exemplified by the pioneering works of Bradley and Bessel.

In parallel were the much larger sky surveys, like those of Flamsteed in the early 1700s at Greenwich, and Lalande in the early 1800s in Paris. For these, the very highest accuracy of individual measurements was sacrificed, and parallaxes were not part of the design. The goal was rather the charting of large numbers of star positions and motions, the motivation being a better understanding of their distribution and their motions through space.

Over the last hundred and fifty years or so, these two branches of star measurements have really split more convincingly into three: small numbers of stars measured with the highest relative accuracy to fix more parallax distances; others spread over the sky and measured with a very good absolute accuracy to give an overall reference frame; and large surveys aimed at elucidating the structure and properties of our Galaxy from the distribution and motion of the stars.

M. Perryman, *The Making of History's Greatest Star Map*, Astronomers' Universe
DOI 10.1007/978-3-642-11602-5_5, © Springer-Verlag Berlin Heidelberg 2010

THE FIRST OF THESE MEASUREMENT BRANCHES focused on a concerted effort to determine more, and more accurate, parallax distances. In the years following the first success of Bessel, initial excitement at the prospect of staking out the space distribution of many more stars was overtaken by the bleak realisation that the majority of bright stars lay at colossal distances that still could not be discerned. Observers had to continue to select target stars fastidiously with the best possible prospects of being nearby, while attention still had to be lavished on a relatively small number of candidates. The measurements remained delicate and time consuming. The highest instrument qualities, meticulous checks for any possible errors, and multiple observations throughout the year were all mandatory.

Creative Commons ShareAlike 3.0

Kuffner heliometer, Vienna

Visual observations using heliometers continued to dominate until the dawn of the twentieth century. A copy of Joseph Fraunhofer's Königsberg heliometer was installed in Bonn in 1848, and a still larger instrument delivered to Wilhelm Struve's group at the imperial Russian observatory in Pulkovo. Others were procured by observatories at Oxford, Stuttgart, Leipzig, Göttingen, and Bamberg in Europe, with the largest such instrument ever made, eight and a half inches in aperture and ten feet long, installed at the Kuffner observatory in Vienna in 1896. David Gill began a heliometer programme in the southern hemisphere at the Cape of Good Hope, and the first in America was started by W. Lewis Elkin at Yale in 1885.

Slowly the number of star distances grew. But progress remained painfully sluggish, and lengthy discussions of the errors reinforced the continuing very great difficulty of the task. Indeed to some it appeared that the era of star parallax measurements was already effectively over; astronomers again, in the words of Hirshfeld, "defeated by the sheer immensity of the realm they were attempting to chart".

What came to the rescue was the new medium of photography. The earliest commercially viable photographic process, daguerrotype, was used by Harvard astronomers J.A. Whipple and William Cranch Bond to capture the first photographic image of the bright star Vega in July 1850. More efficient photographic processes appeared, and early celestial astrophotography by amateur Warren De la Rue in England was followed by the first photographic parallaxes by Charles Pritchard at Oxford in 1886.

Jacobus Kapteyn in Groningen published a list of just 58 parallaxes in 1901. Meridian circles at Leiden and Heidelberg, and photographic plates from Pulkovo and Cambridge, upped the total to 365 by 1910. Yet Kapteyn remained far from satisfied: "Up to the present and for obvious reasons, parallax observers have devoted their labours exclusively to the bright and swiftly moving stars. In our opinion the time has come for a change of tactics. We need the average parallax of the faint stars and of those with moderate and small proper motion as sorely as the rest." His urgent plea was to "extend the investigations into the arrangement of the stars in space."

A new era in photographic parallax determinations was duly opened up by Frank Schlesinger (1871–1943). Astronomy was developing on many fronts, and knowledge of stellar distances became of pressing importance. Schlesinger was born in New York, and his PhD at Columbia University had made use of an unusual benefaction: in 1890, the university had received from the pioneering amateur astrophotographer Lewis Morris Rutherfurd more than a thousand photographic plates of the Sun, Moon, planets and stars taken between 1858 and 1877. Acquired with a thirteen-inch refractor, a particular type of telescope which uses lenses rather than mirrors to focus the starlight, Schlesinger's experience with the plates convinced him that with a high quality telescope of considerable focal length,

Frank Schlesinger

parallaxes could be determined more economically, more conveniently, and more accurately than by any other method.

So it was that at the Yerkes observatory in Wisconsin in 1903, Schlesinger started a parallax programme using their recently completed forty-inch refractor. This giraffe of a telescope, which remains the largest refractor in the world, was designed around a very long focal length to provide the highest magnification of a small carefully-chosen region of the sky, the easier to discern the tiny parallax wobble. Measurements under his direction started at the observatories of Allegheny in Pennsylvania, where he served as director from 1905 to 1920, and were continued by his successors at the observatories of Yerkes in Chicago, Van Vleck in Connecticut, and McCormick in Virginia. His classic papers appeared in print in 1910 and 1911, detailing the results for just twenty eight stars.

Within the next decade, such was his influence, and such was the importance of the task, that eight observatories had made parallax determinations a prominent part of their astronomy programmes.

Wikimedia Commons

Yerkes refractor in 1897

His first task as director of Yale university observatory, a position he held between 1920 and 1941, was to plan a new telescope to further the onslaught. A new twenty six-inch photographic refractor of thirty six feet focal length was designed. This time it was destined for the southern hemisphere, to Johannesburg, where it would carry out for the southern skies a programme similar to that at Allegheny for the north. Schlesinger went to Johannesburg in 1924 to supervise the observatory construction, and its subsequent dedication by the Prince of Wales. At the time of his death twenty years later, a remarkable fifty thousand plates had been exposed, and shipped back to Yale university in New Haven for measurement. From this mountain of glass, a further sixteen hundred precious star distances were distilled. Moved to Australia in 1952 due to deteriorating sky conditions, the telescope was destroyed by a fierce forest fire in January 2003—a tragic ending for an instrument which had pinned down the distances of so many of the brightest stars.

In 1924 Schlesinger published his *General Catalogue of Stellar Parallaxes*, advancing the total known to just short of two thousand, and extending the frail stellar distance network out to a few tens of light-years. His life's work brought him the gold medal of the Royal Astronomical Society in 1927 and the Bruce medal, another of the highest honours in the field of astronomy, in 1929.

For almost a century thereafter, parallax determinations were led by American astronomers. It was said of his methods that they were "basic and complete, and that no major improvements are possible." Grand praise, and no great surprise therefore that almost all other parallax programmes of the same era would follow his approach. Outside the United States, Sir Frank Watson 'six pips' Dyson, Astronomer Royal from 1910 to 1933, and remembered for introducing the iconic time signals in 1924, published twelve years of parallax observations from Greenwich in 1925. His successor as Astronomer Royal, Sir Harold Spencer Jones (1890–1960), published a number of parallaxes of southern hemisphere stars from observations acquired at the Royal Observatory established at the Cape of Good Hope.

At a time when many astronomers were moving to newer fields of astrophysics, a few still dedicated their careers to astrometric measurements of the very highest calibre.

IN ASTRONOMY IT OFTEN HAPPENS that some individual will take the initiative, and rise to the challenge, of making a compilation of all the different work going on around the world in a particular field. With various observatories contributing more distances, often of different quality, and sometimes duplicating attempts at measuring the same star with different instruments, a critical compilation of parallaxes was badly needed. Louise Freeland Jenkins at Yale stepped in to fill a much-needed gap. She brought out a new edition of Schlesinger's *General Catalogue of Trigonometric Stellar Parallaxes* in 1952, with distances for just under six thousand stars based on photographic determinations of the Schlesinger era. A supplement in 1963 raised the total to nearly six and a half thousand.

A further update appeared in 1995. The *Yale Trigonometric Parallax Catalogue* of just over eight thousand stars was pieced together by Yale astronomer William van Altena. It was the catalogue that the world's astronomers consulted near the end of the second millennium if they wanted to know the distance to a star. It was also to be the final collection of ground-based parallaxes before those from Hipparcos.

William van Altena (2005)

At the time of the push to space, the total number of known star distances was certainly respectable, and had been terribly hard won. But even for the population of stars within our solar neighbourhood it was a paltry sampling, let alone viewed in the context of the hundred billion or more stars in our Galaxy as a whole. Crucial and niggling were the plethora of discrepancies and errors arising from the shimmering atmosphere. Accuracies were supposedly around one hundredth of a second of arc, but in reality were often much poorer. This made it difficult for astronomers to rely on published values, and dangerous to draw wide-reaching scientific conclusions. A new approach to measuring distances was sorely needed.

A SECOND MEASUREMENT BRANCH was devouring enormous efforts, in parallel with the work on parallax, to set up the best possible stellar reference frame—to measure and list the positions of a number of agreed reference stars ranged across the entire sky.

To determine a chosen star's distance, repeated positions measured with respect to some other star nearby on the sky would hopefully reveal its parallax motion over the course of a year, but the position of the reference star itself was quite irrelevant. A celestial reference frame demanded, in contrast, a network of precise positions of stars over the entire sky—

a set of agreed reference beacons, with positions and motions well nailed down. Hipparchus and Ulugh Beg, Tycho, Flamsteed and Bradley had typified the earliest efforts to establish a stellar reference system across the celestial sphere.

By the second half of the nineteenth century, a multitude of studies clamoured for a much improved grid of astral trig points. It was needed as a reference frame for the much fainter star surveys starting up to probe the Galaxy's structure, and for studies of the motions of the planets and of the rotation of the Earth. These needs turned to meridian circle instruments to give the best positions for a relatively small numbers of stars. The problem was one that Hipparcos would be well set-up to solve properly later on: that of linking together observations made at different geographic locations and at different times. The reference frame demanded positions of the stars, linked through to the planets, the Moon, and the Sun. A perfidious complication was the fact that the measurement platform, the Earth itself, was slowly 'wobbling' due to effects referred to as precession, nutation, and short-term and unpredictable polar motion.

Portrait: Ernst Hildebrand

Arthur von Auwers

In Germany, a sequence of whole-sky star catalogues, named the FK series after the German Fundamental Katalog, began with the work of Arthur von Auwers (1838–1915) in the late 1870s and early 1880s. Their work was to dominate the field for over a century, although a parallel American effort started with Simon Newcomb's accurate charting of just over a thousand stars in 1899, and continued with Benjamin Boss's influential General Catalogue of 1937.

Auwers had started out on his own career at Königsberg, using Bessel's original heliometer. He made his own measurements of a small number of parallaxes, and established the orbital motion of the binary companion of Sirius based on many thousands of meridian circle observations taken over six years. There were, however, no nearby suitable comparison stars for Sirius, and his experiences in constructing a reference system based upon earlier observations led to the catalogue construction work which would dominate the rest of his life. Auwers began by returning to the very accurate observations made by James Bradley over the years 1750–62, comparing them with more modern observations to determine star motions. This piece of work alone would occupy him from 1866 for a further ten years.

By successive steps, Auwers established a system of just thirty six benchmark stars, with longitudes across the sky fully consistent with each other.

Their origin was set by Bradley's observations of the Sun a century before. Into this he folded other observations, of Bradley's own zenith sector measurements acquired from Greenwich, and others by Nevil Maskelyne around the 1760s and by Stephen Groombridge around 1810. The result was a reference system across the sky of just over three thousand stars. All were reobserved from Greenwich around 1865, to give the most accurate motions of stars to date, pinned down from the grand lever arm of a century and a half of meticulous observation. These motions would form the basis of many pioneering researches into star movements carried out over many decades, including Simon Newcomb's revision of the Earth's wobbling motion, and Jacobus Kapteyn's investigations into the rotating Galaxy. The resulting catalogue was published by the Saint Petersburg Academy of Sciences in 1888. In a later collaboration with David Gill in 1889 to refine the distance to the Sun, Auwers provided observational skills much needed by Gill, which the latter acknowledged with the words at the start of this chapter.

These 'fundamental' catalogues, it should be stressed, charted only a rather small number of reference stars. They gave only one star every six degrees or so on the sky, or just a handful across the whole of Europe if thought of as a mapping of Earth. Successive catalogues added more observations, and slowly yielded a better grid, although rejecting inferior observations also whittled down the number of quality reference stars. More could be interpolated from meridian circle or photographic observations, but inherent distortions would ultimately rest on the quality of the primary grid.

The final catalogue in the series, the FK5, was compiled at Heidelberg under the leadership of Walter Fricke (1915–1988). Fricke had started his career in Hamburg, and worked for a spell at the Yerkes observatory in America. Back in Europe, he worked first in Hamburg, and thereafter at the Astronomisches Rechen-Institut in Heidelberg. There he joined forces with its director August Kopff to derive the fourth fundamental catalogue, the FK4, published in 1963. Thereafter, and as institute director himself from 1954 until 1985, Fricke devoted the next quarter of a century to its upgrade, and its eventual replacement.

The work required to create these catalogues extended over many years of careful observation and critical analysis. The FK5 catalogue was the culmination of a compilation of about 260 individual catalogues, observed mostly with meridian circles and some astrolabes. Like the FK4 it contained just 1535 stars and was published in 1988, just after his death.

Walter Fricke (1985)

ARI archives (Helga Ballmann)

Walter Fricke received widespread recognition for his celestial cartography including, in 1981, the Distinguished Service Cross, First Class, of the Federal Republic of Germany. Beyond his retirement in 1983 he would, in turn, advise the Hipparcos project on its observing programme and, with his successor, would bring his institute's experience in star catalogue production to bear on the space observations.

While individual star positions entering these reference catalogues reached accuracies of several hundredths of a second of arc, and notwithstanding the massive effort and observations invested, evidence still suggested that there were significant hidden errors depending on position on the sky. Years before, Kapteyn said in 1922: "I know of no more depressing thing in the whole domain of astronomy, than to pass from the consideration of the accidental errors of our star places to that of their systematic errors. Whereas many of our meridian instruments are so perfect that by a single observation they determine the coordinates of an equatorial star with a probable error not exceeding two or three tenths of a second of arc, the best result to be obtained from a thousand observations at all of our best observatories together may have a real error of half a second of arc and more."

La Palma Observatory: Dafydd Wyn Evans

Carlsberg meridian circle

Like ancient maps of Earth, the star charts were topologically correct, but stretched and squeezed over the sky in ways that could be guessed but not fully fathomed, hidden errors which proved utterly impossible to track down and remove. They were only fully apparent once the Hipparcos space results were published.

Meridian circles remained the instrument of choice for the highest accuracy surveys until the late twentieth century. Most were phased out after the Hipparcos catalogue was published in 1997, but the automatic eighteen centimeter aperture Carlsberg meridian telescope stands as one of a few exceptions. It was moved to the island of La Palma in 1984, refurbished with a CCD detector in 1998, and continues to operate remotely, turning out more than a hundred thousand star observations each night, their positions locked into the Hipparcos grid.

Walter Fricke's FK5 catalogue of 1535 stars was the final word on the celestial reference frame before the launch of Hipparcos. It was the state-of-the-art in star charting until the space-based positions appeared. But it was far too sparse, and inaccurate, to satisfy modern needs. Like the parallax catalogues, it was impossible to rely on published positions. This was not a good situation for a field so basic. As for the star distances, a new approach to the measurement of a celestial reference frame was required.

THE THIRD MEASUREMENT BRANCH of relevance to astrometry over the last century is represented by the large-scale photographic surveys. This branch traces its roots to the early 1600s, when Galileo used the newly-invented telescope to observe the Milky Way, and found that it could be resolved into innumerable faint stars. By the mid-eighteenth century astronomer Thomas Wright (1711–1786) had described the Milky Way as a flattened disk of stars in which the Sun is itself confined, "an optical effect due to our immersion in what locally approximates to a flat layer of stars."

Philosopher Immanuel Kant (1724–1804) developed these ideas, and also postulated the existence of other 'island universes' distributed throughout space at enormous distances. William Herschel counted the number of stars in different sky regions to deduce the relative dimensions of our Galaxy. These gave valuable insights, but with conclusions founded on the crucial but incorrect assumption that all stars had the same absolute brightness. It was nevertheless becoming clear that large stellar surveys could have much to say about our Galaxy's basic properties such as its size and its shape.

Many large and enormously influential surveys have been made over the last hundred and fifty years. Important amongst the earliest were the huge three-part 'Durchmusterung', named for the German for survey, a word capturing the grandeur of the enterprise. The first two parts were the last of the great star maps to be made visually, pre-dating the use of photography—assistants recorded the positions and magnitudes of stars as the Earth spun and the sky drifted across the fixed telescope field surveying successive latitude zones. The series started with the northern sky surveyed from Bonn by Friedrich Argelander and Eduard Schönfeld. Published between 1852 and 1859, this gave the positions of more than 324 000 stars of the northern the sky. The extension southwards was surveyed from Córdoba in Argentina by John Thome starting in 1892.

The new medium of photography had burst onto the astronomical scene in the late 1800s. Hand-in-hand with the meridian circles giving the highest accuracy reference grid for the brightest stars, photography was to dominate surveys of the skies for the next century. The switch to photography also represented a change in methodology: until then, position measurements had been made by eye, then transcribed to make a star chart. With photography, a chart of the sky was captured directly, and the positions of the stars deduced from them.

David Gill

Lick Observatory archives

Amongst the earliest of these was the southward extension of the Bonn and Córdoba Durchmusterung, covering the southernmost skies from the Cape of Good Hope. The results of the work, led by Sir David Gill (1843–1914) and influential Dutch astronomer Jacobus Kapteyn, were published around the turn of the century. Positions were around one second of arc, limited by the twin barriers of atmospheric turbulence and photographic plate quality. The vast Durchmusterungen triptych was only eventually transcribed to computer form in a fifteen-year effort around the 1980s.

THESE FIRST TRULY LARGE-SCALE surveys provided the foundations on which many later investigations would build their own views of the changing positions of the stars. Thereafter new and deeper surveys from many different observatories around the world contributed to the growing edifice.

Photography allowed the positions of stars to be measured in terrific numbers. With the large telescopes and long exposures of the later 1900s, deep sky images several degrees in extent could yield thousands or millions of star images per plate. The technique was straightforward in principle: exposed at the focus of a telescope tracking the apparent motion of the celestial sky, the plates provided images of stars in large numbers. Their positions on the plates could then be measured and recorded, duly transformed to provide immense catalogues of star positions. In practice, good images require excellent high-altitude observing sites, excellent telescope optics, and accurate and smooth drive mechanisms to track the rotating sky. But they could never eliminate the straightjacket imposed by the atmosphere.

Photographic plates store well for decades, and astronomical libraries and archives across the world preserve a record of how the skies appeared over the past century. As the technology reached its peak in the 1970s and 1980s elaborate and fast automatic measuring machines scanned new and ancient archive plates wholesale. Together they have captured and stored the results in the form of huge digital catalogues of the night sky which will be preserved indefinitely. Yet fundamental distortions due to the telescope optics have always confounded the ultimate accuracies, while the Earth's atmosphere, and the ever-so-slightly dancing images seen through it, still imposes its ever impenetrable barrier.

Although the highest positional accuracies were therefore sacrificed in favour of quantity of stars, massive sky surveys using photographic plates nevertheless changed the course of astronomy. There were various reasons for this impact. First off, simply counting stars to different brightness limits in different directions of the Galaxy has provided many clues as to its structure and dimensions. The technique is especially powerful when interpreted alongside other knowledge, such as the type or temperature of the stars from spectroscopy.

Measuring the same region of sky over many years or decades is particularly effective at revealing the motions of many stars. Photographic surveys, carefully calibrated and repeated decades later, have turned the early detections of star motions by Halley and others into a large-scale discovery factory on an industrial scale. Repeating exposures of the stars over intervals of months or years has another important spin-off: it has led to the discovery of huge numbers of variable stars, their variability over time encoding clues as to their masses, luminosities, and evolutionary states. Star colours measured from different filters and photographic emulsions also provide a wealth of indicators such as their temperature and gravity.

STAR POSITIONS IN LARGE NUMBERS allowed astronomers to embark on a new, more quantitative discussion of our Galaxy's structure. In 1904, studying the Cape Photographic Durchmusterung, which he had worked on in collaboration with David Gill, Kapteyn found that the motions of stars were not random, but could be divided into two streams, moving in nearly opposite directions in different parts of the sky—the first hint of the rotation of our Galaxy. In a summary of his life's work published in 1922, Kapteyn described the Galaxy as a lens-shaped island universe in which the density of stars decreased away from its centre. In his model, the Galaxy was held to be some 40 000 light-years in size, not so far from our present ideas. But, as if clutching at long-held belief that the Earth must occupy some privileged place in the Universe, Kaptyen held that the Sun was relatively close to its centre, at around two thousand light-years.

Jacobus Kapteyn

The size of the Galaxy, and the distance scale within it, became issues of great debate. It was not easy to infer the structure of the Galaxy from star counts alone, and there were many complications. Great clouds of dense interstellar gas occupy various pockets within our Galaxy's disk, and these block out the more distant light from stars beyond. It's not so different to looking at the night sky covered by thin cloud. With no simple means to identify the gas, seeing only a few stars along a particular sight line might suggest that the Galaxy was only thinly populated by stars in that direction, while the very opposite might be true. Another tricky problem was caused by the growing realisation that stars were of many different types, with hugely varying luminosities and very different types of motion through space. So evident in retrospect, trying to figure out the properties of our Galaxy from an erroneous census was doomed to fail.

So it was that even into the 1920s, the detailed structure of our own Galaxy, and the relationship between it and those that we now know lie far beyond, still remained a puzzle. The uncertainties precipitated an exchange which has gone down as astronomy's Great Debate, which took place on 26 April 1920 in the Smithsonian Museum of Natural History in Washington DC. Harlow Shapley, from the Mount Wilson observatory, argued that our Sun lay far from the centre of a single Great Galaxy, in which spiral nebulae such as Andromeda were simply part of our own. Heber Curtis, from the Allegheny observatory, disagreed. He held that the Sun was near the centre of a relatively small Galaxy, with the entire Universe composed of many other galaxies somewhat like our own. It was a debate deeply rooted in the uncertainty of the scale of the Universe which had still not been resolved.

Edwin Hubble's identification of pulsating Cepheid variables in the Andromeda nebula in the mid-1920s confirmed that it was a distant galaxy much like our own, but far beyond. Like brilliant lighthouses pulsing across the depths of space, these standard candles illuminated our understanding of the scale on which the Universe is constructed. Shapley was duly proven more correct about the size of our Galaxy and the Sun's location within it. But Curtis's view that the Universe was composed of many more galaxies, and that 'spiral nebulae' were indeed galaxies just like our own, was corroborated. With almost a century's hindsight, the debate is important[26] "not only as a historical document, but also as a glimpse into the reasoning processes of eminent scientists engaged in a great controversy for which the evidence on both sides is fragmentary and partly faulty."

SCHMIDT TELESCOPES APPEARED on the scene in the second half of the twentieth century, and brought their own revolution. Named after their optical designer Bernhard Schmidt, a cleverly-designed corrector lens positioned in front of the primary reflecting mirror resulted in unprecedentedly large fields of view of several degrees on a side. This made it possible to observe a still bigger patch on the sky, several times the diameter of the full Moon, in a single exposure. As a result, Schmidt telescopes contributed a flood of high-quality observations that brought positional astronomy back to the fore.

Monumental surveys were carried out from Palomar Mountain in California from 1949, in a grand programme funded by a grant from the National Geographic Society to the California Institute of Technology. The southern skies were surveyed by the European Southern Observatory from the mountain top observatory on La Silla in Chile from 1973, and from the UK's observatory in Australia from about the same time.

The surveys produced thousands of meticulously exposed plates which were themselves reproduced photographically, and circulated in limited editions to the world's astronomical institutes for detailed scrutiny. Collectively, they comprise hundreds of billions of star images, an archival view of the celestial sky as it will never be seen again. The resulting vast catalogues, of more than a billion stars across the sky, are used for countless astronomical projects, includ-

UK 1.2-meter Schmidt telescope

ing pointing their way around the sky by the great space observatories, the Hubble Space Telescope amongst them.

Photographic plate surveys made far in the past—a century or more ago—remain of great value to present day astronomy, for a repeat survey today will easily identify the most rapid movers with the largest motions. Catalogues of stellar motions continue to be constructed from various combinations of these photographic plates, using the same technique which allowed Edmond Halley to identify the first stellar motions three hundred years ago.

Many of these grand twentieth century photographic surveys have been revitalised by the results of the Hipparcos space mission. The new reference system from space can be propagated backwards in time using the measured proper motions, to give an improved reference system for the years that the plates were taken. The improved reference system then gave much better positions for the large numbers of other stars on the plates. This, in turn, has led to vastly improved star motions tracked between the times of the earliest photographic plates a century ago, and the measurements from space made in the last decade of the second millennium.

IN THIS CONTEXT, ONE REMARKABLE project deserves specific mention: the imposingly named *Carte du Ciel*, the Map of the Heavens. This vast and unprecedented international star-mapping project was initiated by ex-naval officer and Paris Observatory director Rear Admiral Amédée Mouchez, in collaboration with Sir David Gill, Her Majesty's Astronomer at the Cape of Good Hope at the time.

Amédée Mouchez

Mouchez had started his career with hydrographic studies along the coasts of Korea, China and South America and later, during the Franco–Prussian War, led a heroic defence of Le Havre. Taking the helm at the Paris Observatory, correspondence between Mouchez and Gill led to the "assembling of a great international conference", the Astrographic Congress of more than fifty astronomers held in Paris, on 16 April 1887. Participants included Auwers from Germany, Kapteyn from The Netherlands, Struve from Russia, and William Christie, the Astronomer Royal from England.

The new medium of astronomical photography offered a remarkable possibility to carry out a celestial survey totally unprecedented in the history of astronomy, and astronomers seized the opportunity. The objectives of this first ever international astronomical collaboration on a massive scale were intoxicating but would prove to be overwhelming. The idea was to build up and deploy a system of identical telescopes straddling the full range of latitudes on Earth, survey the sky, and build up a monumental star catalogue as a result. According to H.H. Turner in his description of the project[27]: "The discussions were, to say the least of it, animated. There are no universal rules for conducting public business, and astronomers from one country were not familiar with rules in use elsewhere. It interested Englishmen, for instance, who are accustomed to have resolutions moved by anyone rather than the chairman, to learn that this was by no means a universal rule. On the contrary the chairman of the first conference considered it part of his duties to move all the resolutions. After listening to a discussion, he took it to be his function to summarise the sense of the meeting in a resolution which he put from the chair and in favour of which he held up his own hand. Unfortunately for his success his was sometimes the only hand held up, and the discussion was necessarily resumed."

Turner considered that the conference was: "...a remarkable meeting, the first of its kind in the history of astronomy; and it has shown the way for subsequent gatherings... On all of these occasions the French have acted as hosts and have discharged these duties with a cordiality and hospitality that has never failed to impress their colleagues from the most distant parts of the world."

The ambitious enterprise had two separate yet connected parts. The first, the Astrographic Catalogue, would photograph the entire sky to 11 magnitude, thereby picking out stars a hundred times fainter than the feeblest seen by the unaided eye. It would provide a plentiful reference catalogue much denser than anything observed by transit instruments.

Twenty observatories around the world participated, each choosing a strip of sky convenient in latitude. Each would procure the necessary telescope, suitably equipped and staffed. Then collectively they would expose, for six minutes each, more than twenty thousand glass plates of the night sky. Turner estimated the total weight of these plates at three tons.

A key agreement, and one essential to the survey uniformity, was to use similar telescopes. Around half of the observatories eventually procured astrographs from the Henry brothers in France, with the others coming from the firm of Howard Grubb in Dublin. The different observa-

Greenwich astrograph (c. 1920)

tories were assigned different latitude strips to photograph: Greenwich, the Vatican, Catania, Helsing, Potsdam and Hyderabad would cover the northern sky. Uccle, Oxford, Paris, Bordeaux, Toulouse, Algiers, San Fernando and Tacuba would span the equatorial regions. Córdoba, Perth, Cape of Good Hope, Sydney, and Melbourne would survey the southern skies.

The first plate was taken in August 1891 at the Vatican Observatory. The exposures there, taken by the hands of a single observer, took more than twenty seven years to complete. The very last plate was finally exposed in December 1950 at the Uccle Observatory in Bruxelles.

The plates were in due course photographed, measured, and the results published in their entirety, providing star positions with an accuracy of about half a second of arc. In practice, the measurements were a highly protracted affair, with the tasks around the world assigned to willing—and in some cases unwilling—assistants[28]. All measurements of the star images were made by eye, and recorded by hand. In many observatories—Paris, Melbourne, Perth, Cape, Toulouse and others—twenty or thirty women (the original 'computers') were taken on to help with the herculean task. For the Vatican

Nuns measuring the Vatican zone Carte du Ciel plates (c. 1900)

plate collection, archival photographs show nuns from the Congregation of the Child Mary at work measuring the plates. Turner commented that "each observatory has thus to measure about half a million star images… These measures took a staff of four or five people at Oxford some ten years or so to complete: and the printing of them another four years."

In total, nearly five million stars were recorded. Publication of the various parts proceeded from 1902 to 1964, and resulted in a massive two hundred and fifty four printed volumes.

For the second part of the conference goals of 1887, a further set of plates, with longer exposures but minimal overlap, would photograph all stars to 14 magnitude—the magnitude scale being logarithmic, this corresponds to stars a thousand times fainter than those that can be seen with the naked eye. Most of these plates used three exposures of twenty minutes each, displaced to form a small triangle with sides of ten seconds of arc, making it easier to distinguish stars from plate flaws, and to differentiate stars from the more rapidly-moving asteroids. The grand idea was that exposed plates would be reproduced and distributed as a set of charts, the *Carte du Ciel*. However, reproduction of the charts, originally to be undertaken using engraved copper plates, proved to be prohibitively expensive, and many zones were either not completed or not properly published.

Despite, or perhaps because of, its vast scale, the project was only ever partially successful, even though many commited individuals had devoted tedious decades of their careers to its success. The *Carte du Ciel* component was never completed, and the Astrographic Catalogue lay largely ignored for nearly a century. Its star positions were difficult to work with because they were not available in computerised form, and neither were they listed in convenient coordinates. Some historians of science have classified this vast project as the story of how the best European observatories of the nineteenth century lost their leadership in astronomy by committing vast resources to a somewhat misguided undertaking.

Long portrayed as an object lesson in over-ambition, languishing lost and forgotten for a century, the Astrographic Catalogue made a remarkable reappearance on the world's astronomical stage barely a decade ago. The Hipparcos catalogue positions could be used, in combination with each star's proper motion, to provide a reference frame back at the time when the Astrographic Catalogue plates were taken. So calibrated, they gave the places of all catalogue stars which they occupied in the sky some one hundred years before. Combining those with the satellite positions nearly a century later gave extremely accurate motions for two and a half million stars.

Like the ancient catalogue of Hipparchus dusted off and used to reveal star motions by Halley, the Astrographic Catalogue is a remarkable example of an all-but-abandoned project, for whom so many had toiled for so long, waiting patiently to prove its inestimable value generations afterwards.

IN THE LAST TWENTY YEARS, photographic plates have all but disappeared from astronomy. They have gone the way of sextants and quadrants and most meridian circles before them. In their place the CCD, the ultra-sensitive solid-state silicon detectors, of the type used in digital cameras and video recorders, has taken over the challenge, and has brought with it another revolution in surveying the skies. The full-sky surveys of the US Naval Observatory based in Washington DC, and the Sloan Digital Sky Survey supported by the Alfred P. Sloan Foundation (a philanthropic structure set up by the one-time President of General Motors), have led this new wave, leading to deeper exposures, and more stars, than ever before.

Other magisterial CCD surveys are even now coming on line, from high-altitude sites in the Atacama desert and perched in the mountain top observatories of Hawaii. The emphasis has evolved somewhat, to surveying the sky as quickly as possible in as many colour filters as technically feasible. New challenges come as these unprecedented surveys scan the night skies, over and over, with a speed and sensitivity inconceivable only a couple of decades before.

The data flood will hit astronomers like a tsunami, bearing all sorts of new messages about the Universe, and the challenge will again turn to interpreting what messages these new observations encode.

A FINAL MIX OF CURIOUS PHENOMENA showed up in the measurement of the accurate positions of the stars and the planets over the last couple of centuries, bringing us back, in full circle, to the earliest of the Greek studies of the fixed stars and the wandering planets.

Objects in our daily lives are generally not massive enough, or the effects not measurable accurately enough, for Newton's Law to be examined for real flaws or imperfections. But the motions of the planets provide a miraculous laboratory for observing the most delicate touches of gravity. Alongside innumerable other successes of Newtonian gravity was its part in the discovery of the planet Neptune.

Meudon Observatory, Paris (2002)

In the middle of the nineteenth century French mathematician Urbain Le Verrier (1811–1877), working under François Arago at the Paris Observatory, had been making a careful study of the orbit of Uranus. There were small but systematic discrepancies between its observed orbit, and that predicted by Newtonian theory—its measured position was consistently off from where theory forecast it should be. Something was wrong.

Newtonian gravity had proven itself repeatedly and was not the suspect. Le Verrier was forced to conclude that an undiscovered planet existed out in the far reaches of the solar system, giving erratic tugs at Uranus during its journey around the Sun. He could predict a position for an unknown object which, he believed, must be responsible for disturbing its orbit. Neptune, as it would be called, was duly discovered by Johann Galle and Heinrich d'Arrest, within one degree of his predicted location, on 23 September 1846. It was a triumph for Newtonian gravity, and a sensational result for Le Verrier, who became director of the Paris Observatory in 1854, following in the footsteps of Cassini and Lalande. A source of debate ever since has been the extent to which John Couch Adams, who had made similar calculations even earlier, should also be credited with Neptune's discovery.

The earliest and most worrying sign that all was not completely well with Newtonian theory was the detailed motion of our innermost planet. Mercury circles the Sun in a tight, bakingly-hot elliptical orbit of just ninety days. Its point of closest approach advances around the Sun by a small amount each year, about one minute of arc, due to various effects, including the gravitational pull of the other planets.

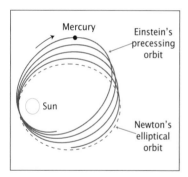

Precession in the orbit of Mercury

Le Verrier noticed that the slowly changing shift could not be explained completely by Newton's laws. There was a tiny mismatch of a little less than half a second of arc per year, an almost undetectable amount, except for the fact that it rolls up and accumulates with time, to nearly forty three seconds of arc each century. In 1843, inspired by his success with Neptune, Le Verrier published his interpretation of the mismatch as being due to a hypothetical inner planet, which he named Vulcan. This precipitated a search for the new planet, and a wave of false detections that would follow unabated over the next sixty years. One Edmond Lescarbault was even awarded France's prestigious Légion d'honneur for his claimed sighting of the non-existent body.

In 1915, while the searches were in full swing, Albert Einstein published his general theory of relativity. This describes gravity as a basic property of the geometry of space and time, a distortion in their very fabric due to the presence of mass. It superseded Newton's law of universal gravitation as 'the' theory of gravity. Mathematicians like its elegance, and physicists like it because it gives hints as to why this force exists. Mostly the predictions of Newton and Einstein agree. But in certain situations they differ, slightly but significantly, and tests to confirm or repudiate it were eagerly sought.

The orbit of Mercury was an obvious target. It was Einstein himself who showed that his theory explained exactly the discrepancy, important evidence that he had identified the correct form of the equations describing gravity. The effect, grandly referred to as perihelion precession, has also been seen for Venus and Earth. In a very close binary pulsar system, discovered in 1974, the effect is an astonishing hundred thousand times larger. In all cases, theory and observation are in precise accord.

Albert Einstein

Le Verrier died in 1877 still convinced that he had detected a second planet. Yet while most of the interest in Vulcan naturally evaporated, claims and counterclaims of asteroid transits, and searches for Vulcanoid asteroids orbiting close to the Sun, still continue to the present day.

ANOTHER TEST PROVED to be still more compelling. In science a theory that explains a known discrepancy, like Mercury's orbit, is one thing. A specific prediction of a hitherto unknown consequence of a theory is quite another; a wager on which reputations ride high.

According to the prescriptions of general relativity, starlight should be deflected by a very tiny but entirely predictable amount as it passes from a distant star close to the limb of the massive Sun on its way to an observer on Earth. The size of the deflection was predicted to be very small, just over one second of arc at the very limb of the Sun where the effect would be largest. Barely at the limit of the dancing motion of the atmospheric ripples, it would demand careful measure, and an excellent knowledge of the undeflected star image positions to compare with.

It would be impossible to measure position shifts of faint stars close to the limb of the brightest object in the entire sky except, perhaps, if they could exploit the exceptional conditions of a total solar eclipse. This was American solar astronomer George Ellery Hale's proposal to Einstein when asked to suggest an appropriate test. A German–USA expedition planned for an eclipse passing over Crimea in 1914 was foiled by the outbreak of war. The first observations of this light bending were eventually made during the total eclipse of 29 May 1919. Astronomer Royal Sir Frank Watson Dyson had identified this as an auspicious celestial alignment because the Sun and Moon would pass in front of the bright

Frank Watson Dyson

Hyades cluster, more bright stars making it easier to detect changes in their

position. The undeflected star positions that would later be observable close to the Sun's limb during the eclipse had been observed six months previously by night.

Sir Arthur Eddington and Edwin Cottingham from Cambridge journeyed to the West African island of Príncipe in the Gulf of Guinea, while Andrew Crommelin and Charles Davidson from the Royal Greenwich Observatory set up their base near the Brazilian town of Sobral—the two observing stations chosen to improve prospects of observing the eclipse in case of poor weather. During the eclipse, as the sky was plunged into darkness, a few bright stars popped into view and remained visible for two or three minutes. This time, their positions would be minutely deflected by the presence of the Sun's huge gravitating mass reposing squarely along the light path from the distant stars behind the Sun to the observers on Earth.

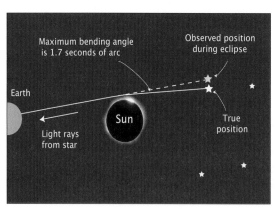

Light bending by the Sun (exagerrated)

The agreement between the small extra shifts observed on the one hand, and Einstein's theory on the other, was very much at the limit of star measurement accuracies of the time. Confirmation of the predicted bending was duly claimed, and widely greeted as spectacular news. It made the front page of major newspapers, making the theory of general relativity world famous, and Einstein himself even more so. When asked what he would have said had his theory not been proven by the observation, Einstein notoriously replied "I would have had to pity our dear Lord. The theory is correct all the same."

Debate about the quality of these early observations has continued, but the theory itself is now unquestioned. Better measurements for other solar eclipses, including one in June 1973 by Texan astronomers from a desert site near Chinguetti in Mauritania, sightings of quasars at radio frequencies, gravitational lenses observed in astronomy in the 1980s, gravitational redshift as perfectly accounted for by GPS navigation satellites, and many other subtle manifestations, have confirmed general relativity as our best description of gravity to date. And Hipparcos would add its own precise measurements of light bending in due course, mapped from the pattern of more than a hundred thousand star ripples across the entire sky.

A RETROSPECTIVE OF OUR LENGTHY journey through history is in order. Two thousand years of charting the stars has led us on a remarkable voyage of discovery. The Earth, as we now know, is not at all at the centre of the Universe, but a spinning rock which orbits the Sun. Billions of other stars, as well as planets, interstellar gas and dust, radiation, and invisible material are bound together to form our Galaxy—a magnificent disk spiral system, prevented from collapsing by its own rotation. Our Sun lies way out in one of the spiral arms, thirty thousand light-years from the centre. Around us the stars, at truly immense distances, move along their own eternal paths. Beyond our own island universe, the Milky Way, a seeming infinity of other galaxies recede from us at astonishing speeds, pointing their fingers backwards in time to the dawn of creation.

Many of these advances in our understanding have accrued from a steady refinement in measuring star positions. Over the past century, improvements advanced along a very high accuracy branch for a very few stars, culminating in the compilations of parallax distances for around eight thousand stars. A medium accuracy branch for a thousand or so stars gave our very best, but still troublingly inadequate, celestial reference system.

The lower accuracy branch developed progressively from Tycho's catalogue of a thousand stars with an accuracy of fifty seconds of arc in around 1600, Flamsteed's survey of three thousand stars to twenty seconds of arc around 1700, Lalande's fifty thousand stars at three seconds of arc around 1780, and Argelander's survey of more than three hundred thousand stars at one second of arc around the 1850s. Billion star surveys were compiled from the world's arsenal of Schmidt telescopes in the late 1900s, but despite their colossal strength in numbers, their positions were only marginally better than the celestial surveys of more than a century before.

At the dawn of the third millennium, the quality of star positions lagged far behind the progress chalked up in many other areas of astronomy. Accurate distances were still only known for a few hundred nearby stars, a severe barrier to progress in understanding the physical processes within the stars. Accuracies from the large photographic surveys were strongly limited by the atmosphere. Proper motions were known for millions of stars, but with distortions over the sky which confounded their interpretation. Distance information needed to transform them to space motions was all but lacking.

PUBLICATION OF the Hipparcos catalogue in 1997 would bring together these various branches, giving positions, space motions, and distances of more than a hundred thousand stars, all measured with equal attention, all accurate to around one thousandth of a second of arc. The catalogue would not only be the most accurate positional survey to date by far. Very significantly, it would join together in a single survey the most delicate work on individual stellar distances, the highest accuracy of the best reference frames, and the formidable large-scale surveys of history's great star charts.

Its massive leap in accuracy would be the largest single advance ever made, an improvement over its predecessors by a factor of fifty. Such a jump had never been witnessed before in the entire history of the field. It would all be done with a small telescope in space, launched on a giant rocket. It would all be done, one might say wryly, with lots of smoke and a few highly-polished mirrors.

Chapter **5**

The Push to Space

*An association of men who will not quarrel with one another is
a thing which never yet existed, from the greatest confederacy of
nations down to a town meeting or a vestry.*

Thomas Jefferson, Letters (1743–1826)

B Y THE SECOND HALF of the twentieth century the steady advance in the
accuracy of stellar positions was running headlong into a number of
essentially insurmountable barriers. Progress in telescopes and their instruments seen over the previous two or three centuries was running out
of steam. Limited improvement in measuring star positions, in turn, obstructed further progress in fixing star distances and studying their space
motions. For once, telescope size or optical quality were no longer the limit-
ing factors. After two millennia of hard-won improvements, human ingenuity appeared to be finally barred by Nature's innate complexity.

The biggest problem was the bending and twinkling effects of the atmosphere, condemning star images to their eternal and unpredictable wobbling dance. New thin-mirror technologies were having some success in correcting effects over small angles, but all attempts to nail down large angles
across the sky failed miserably. In addition, there were the tiny variations in
telescope alignment as the mountain-top observatories went through their
endless day and night cycles of warming and cooling. The variable flexing
of telescopes under their own weight as the huge supporting structures were
steered to observe different parts of the sky added other unpredictable distortions.

M. Perryman, *The Making of History's Greatest Star Map*, Astronomers' Universe
DOI 10.1007/978-3-642-11602-5_6, © Springer-Verlag Berlin Heidelberg 2010

Yet another unassailable complication was that any telescope on Earth can observe only part of the sky at any one time: a telescope in the northern hemisphere only ever sees the northern skies. Even so, it still requires a year to elapse for the entire region to be observable by night. A grid of star positions spanning the entire sky could only be constructed from a vast spider web of thousands of geometrical triangulations from separate telescopes observing accessible portions of the sky at different times. However, between the various observations which had to be carefully patched together, all of the star images had moved, all but chaotically, by the tiny amounts which were to be probed.

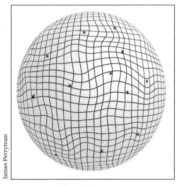

Warped grid in a star map

Like an ancient cartographic survey of the Earth made with primitive surveying instruments, the result of centuries of effort was a map of the sky of sorts, but one which was highly distorted and unpredictably warped. At accuracies below a second of arc, it was simply unreliable. Star positions were plagued by unfathomable errors which could not be unravelled. Their space motions were, in consequence, of variable and sometimes questionable quality. More importantly, distances remained largely unknown, the tiny signatures of their miniscule parallaxes buried under a shroud of error-prone measurements imposed by the flickering atmosphere. A fundamentally new approach to measuring star positions was desperately required.

THE PRECOCIOUS PROPOSAL to make these delicate observations from space was the next master stroke of instrumental creativity. It was first formally laid out in front of other scientists in the mid-1960s by 61-year old French astronomer Pierre Lacroute, although the idea of astrometric detection of binary stars and planets from space had been aired by Paul Couteau and Jean-Claude Pecker in a restricted bulletin of the Nice Observatory a couple of years before that. Until then space science, still very much in its first flush of youth, had been somewhat the preserve of magnetospheric experts studying the region of the Earth's environment controlled by its magnetic field, discovered by Explorer–1 in 1958. X-ray astronomers, meanwhile, were eagerly following up their discovery of the first cosmic X-ray source in 1962.

It seems even more remarkable in hindsight that such a specialised goal in space science should have followed so closely on the heels, within just a decade, of the first ever artificial satellite to orbit the Earth, the Soviet Union's Sputnik 1 in 1957.

Lacroute's compatriot Pierre Bacchus, assistant professor at the Strasbourg Observatory at the time, and later director of the Lille Observatory, was also involved in these early discussions. But the idea of dedicating an expensive space platform to measure star positions came out of left field and was, for a number of years, neither enthusiastically received nor widely embraced. Beyond a limited peer group of its active exponents, the proposal most likely appeared to be a misdirection of the limited opportunities of space funding. Seen from outside, it probably didn't seem too difficult to tighten up the measurements of star positions made from the ground; other people's problems are, after all, never quite as taxing as one's own.

Astrometry was also the victim of its own difficulties. Despite the innovative efforts of brilliant instrumentalists, substantial progress had been painfully slow because the problems were so forbidding. As a result, exciting new scientific results flowing from their work had largely dried up, and new creative minds were ill-inclined to enter the field. From the outside, the discipline probably appeared to be one which had run its course. Wide support for the funding of an expensive space mission would be all the more difficult to engender.

Director of the Strasbourg Observatory for thirty years until his retirement in 1976, his life's work devoted to the measurement of star positions, Lacroute had realised that a space telescope would allow the measurement of arcs and triangulations to be made above the flickering effects of the atmosphere. Also, beyond the buckling forces of Earth's gravity, the telescope would not be sagging unpredictably as it made its cosmic census.

Early discussions, Bordeaux, October 1965
(Lacroute seated at front, Bacchus behind)

Far from the Earth, the satellite would have an uninterrupted view of the entire sky, and the experiment could also be shielded to simulate perpetual night time. It may seem strange to think of space being so dark, but the back side of a satellite shields the telescope from the Sun in just the same way as the Earth shields us at night. In fact, the shielding is much better because there is no atmosphere to scatter sunlight back into the telescope. So from a satellite above the Earth, the skies are very dark indeed. There would always be a region in the direction of the bright Sun which could not be observed at any one time. But even this gap would be filled in after some weeks, as the Earth moved on in its annual orbit around the sky.

Pierre Lacroute (1993)

CNES

The most ingenious part of Lacroute's idea, however, was to observe in two very widely separated directions at the same time. Combining these two different sight lines into a single telescope focus, by means of a special split mirror looking out in two directions simultaneously, would give a network of wide-angle measurements spanning the whole celestial sphere in its entirety.

The idea of making differential angular measurements was not new in itself, and indeed Friedrich Bessel's first parallax measurements had made use of a somewhat similar approach a century and a half before. The novelty, empowered by the elimination of the atmosphere, was making these differential angular measurements across very wide sweeps of the night sky. From the network of space measurements, strict trigonometric distances could be disentangled. The goal, in short, was to construct a vastly improved census of stellar parallaxes, so that their distances could be measured and their physical properties derived. The satellite concept was duly named Hipparcos, a somewhat contrived, and thereafter rarely used, contraction of 'high-precision parallax collecting satellite', but also paying tribute to the ancient Greek pioneer of celestial mapping.

Careful mathematical studies and computer simulations showed that the celestial survey would be both superbly accurate and immensely rigid. It would be like replacing a highly-distorted map of the world made from sailing ships in the 1500s, with one established by satellite imagery and GPS positional technology in the year 2000.

LACROUTE'S FIRST PROPOSAL for these space observations was presented at the thirteenth triennial General Assembly of the world's astronomers, the International Astronomical Union, held in Prague in August 1967. There is a time-honoured way in science for gathering ideas to improve on any new experiment, and that is to publish an early concept, and wait for others to circulate their own criticisms or suggestions for improvements. The following years accordingly saw Lacroute's ideas becoming increasingly developed within the active French community. But beyond France, while the grand vision of space astrometry attracted interest, some of his central ideas were considered technically unrealistic. It would take almost another decade, and the involvement of other talented instrument designers, to transform the early concepts into something truly feasible.

Lacroute had originally hoped to have his ideas supported and duly carried into space by the French national space agency, CNES. France, as some of the other larger European countries, like Germany, Italy, and UK, had its own vibrant space programme which, before and since, has chalked up a string of impressive scientific, telecommunications, and Earth observation satellites. But their studies of the early satellite concept suggested that it would be too complex, too expensive, and too risky for France to go it alone. In particular, Lacroute's proposed implementation of the beam-combining mirror which would, with unprecedented acuity and supreme stability, look out in different directions at the same time, was quickly found to be a critical and formidable technical challenge.

Pooling financial resources, management experience, and technical know-how, the European Space Agency (then ESRO) provided the next obvious choice to the growing lobby for space astrometry for a possible funding and development route. French astronomer Jean Kovalevsky, an early supporter of Lacroute's idea, played a key part in converting the project into a wider European venture.

BACK IN 1960, an intergovernmental conference at Meyrin, in Switzerland, had agreed on setting up a pan-European preparatory commission for the coordination of space research. At a 1962 meeting in London, delegates from Belgium, France, Germany, Italy, The Netherlands, and the United Kingdom signed a convention creating the European Launcher Development Organisation, ELDO. Later that year in Paris, delegates from the same countries, along with Denmark, Spain, Sweden, and Switzerland, agreed to the creation of the European Space Research Organisation, ESRO.

ESA itself was formed in 1973, when the European Space Conference meeting in Brussels decided on the merger between these two bodies—ELDO and ESRO—as well as on the start of the Spacelab and Ariane programmes. Today, ESA is an organisation of eighteen member states which participate to varying degrees in a whole range of mandatory and optional space programmes, across several disciplines or directorates.

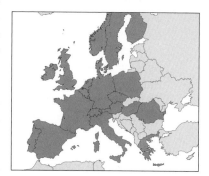

European Space Agency, 2009 (including cooperating states)

Opening up the Hipparcos project to a wider body of international collaborators was to prove enormously beneficial. The best of all interested scientists from across Europe could contribute. Those with creative experimental insight added important new features to the instrument design, and others assisted with developing the astronomical arguments necessary to gain the support of the wider scientific community. Some would embark on a marathon journey to devote years of their lives on figuring out how the data would be analysed and interpreted once beamed down from space.

By the end of the 1970s, Hipparcos looked very different from the early ideas laid down by Lacroute. A symposium organised by ESA in October 1974 in Frascati, Italy, tested its wider international interest, and confirmed its potential pan-European appeal. With a number of crucial improvements brought by experienced Danish astrometrist Erik Høg after his introduction to the project in 1975, the satellite had been transformed. It was yet bolder in its objectives, far more efficient, technically simpler, and therefore more feasible. As a result, it had also succeeded in generating a substantial scientific following across Europe, backed by an increasingly vocal international community. After a decade of study, it was at last poised to compete with other proposals for space missions vying for ESA's approbation and acceptance.

The project always fell somewhat short on instant glamour. But it garnered wide international acclaim for its profound scientific value. Its lack of allure beyond the immediate specialists would even prove to be a trump card during its execution: it was a scientific vision ignored by the other space agencies and, in consequence, encountered no competition during its development. Europe would duly win the unspoken race to measure the stars from space, a victory clearly facilitated by the absence of other competitors.

THE TRANSITION FROM INITIAL CONCEPT to its acceptance within ESA's scientific programme in 1980 followed a precarious path. For any major space facility the process involves considerable preparation and lobbying, and acceptance or rejection usually follows a long and protracted course through the relevant advisory committees.

Science is just one of several broad topical areas of the European Space Agency: others include telecommunications, meteorology, Earth observation, and human space flight. Each has its own advisory committees and decision-making structures. Then, as today, ESA solicited ideas for new missions, studied them intensively, consulted widely, evaluated competing concepts, and duly ruled on the next experiment to be flown in space under the flag of Europe.

The goal was and is still to select, once every few years, ideas which are innovative and scientifically compelling, challenging enough to develop European industrial capabilities while still being technically feasible, and financially acceptable. The competition is always fierce and keenly fought—projects can be studied intensively over a number of years, only to lose out in the face of competing projects as they advance over and around the progressive hurdles of the Agency's advisory bodies. The serious decision-making process kicks in once a set of missions proposals have been studied, and once a set of well-structured science goals and design concepts are on the table.

Within the scientific programme, a hierarchy of external advisory and decision-making structures formulates European space science strategy and policy. At the top, wielding political and financial muscle as well as the ultimate authority of scientific endorsement or veto, the Science Programme Committee bridges national and European interests. Advising it is the influential Science Advisory Committee, which itself bestrides two further subordinate committees: the Solar System Working Group and the Astronomy Working Group. The deliberations of these two groups also aims to create some balance in ESA's science programme between space missions focused respectively on the solar system (the Sun, the Sun–Earth system, the inner and outer planets, their moons, asteroids and comets) and the stars and galaxies beyond (as observed by optical, infrared, ultraviolet and X-ray instruments). All of these committees, including their respective chairs, are populated by scientific advisors external to ESA. Their members are appointed from universities or technical institutes across the various countries to provide objective but in-depth guidance.

The selection process works as follows: the two working groups look at new ideas within their specific area of expertise, and make the first scientific assessment. Their recommendations are passed up one level to the Science Advisory Committee. Smaller in composition, typically more senior in career progression, and expected to be correspondingly more dispassionate in their deliberations, this committee then has to tread the difficult line of weighing and arbitrating across these two somewhat disparate disciplines. Their mandate is to consider the importance of the project's scientific aims, rather than political or financial issues, although the triplet of constraints are closely welded, and difficult to completely separate. After due deliberation, the Science Advisory Committee provides its scientific advice to the Science Programme Committee, the Agency's most senior external science advisory body.

This final committee stage has the additional delicate task of combining impartial scientific guidance with technical, political, and financial considerations, as well as overall programme balance and national interests. The ingredients, once stirred, can occasionally make for an explosive mix. This senior body is accordingly comprised of all the ESA member state national delegations, where senior scientists and space policy makers sit side by side under the chairmanship of one of the elected member state delegates.

ESA headquarters, Paris

Their formal and decisive meetings are held in ESA's headquarters in Paris three or four times a year, with interventions translated simultaneously into English, French and German, as laid down in its constitution. Through the national delegations, the Science Programme Committee in turn advises ESA's Director of Science as to how it wishes the scientific programme to be executed. During the 'Hipparcos years', from 1980 to 1997, the post of Director of Science was occupied for the first three by German Ernst Trendelenburg, and thereafter by French solar physicist Roger–Maurice Bonnet.

THIRTY YEARS ON, the main protagonists probably have somewhat different recollections of the most crucial steps that were trodden as Hipparcos inched forward. However, the main decision points in its early path have been documented in the authoritative history of the European Space Agency[29], from which I quote here *in extenso*.

Following earlier recommendations of the advisory structure in the mid-1970s, a number of competing ideas for a new round of space missions had been studied in 1977 and early 1978, aiming for a final decision in 1979. At the first committee stage, the Astronomy and Solar System Working Groups examined the ideas that had been put forward for new missions in their respective areas of expertise.

Hipparcos was one of the two competitors placed in front of the Astronomy Working Group, the other being a telescope to probe the Universe at ultraviolet wavelength, EXUV. The committee eventually awarded its priority to astrometry. The voting was a reasonably unambiguous eight votes to three, so for Hipparcos, so far so good.

The Solar System Working Group, meanwhile, had to rule on a competition between POLO, a proposed polar orbiting lunar observatory, and Giotto, the mission being proposed to rendezvous with Comet Halley. The struggle was hard fought, but Giotto duly emerged triumphantly as the committee's unequivocal choice. The recommendation also came with its own certain frisson: the Giotto spacecraft[30] would have to be developed and launched with an urgency and speed unprecedented within European space science, with launch just five years hence. Only on this aggressive schedule would there be time for it to race to its rendezvous with Comet Halley which, in its seventy-six year orbit around the Sun, would make its final apparition of the millennium in 1986.

When the Science Advisory Committee assembled for its meeting on 6–7 February 1980, the choice before them was therefore between Hipparcos and Giotto. Astrometry of the stars or rendezvous with a comet? The aspirations of the astronomy community, or the wishes of the solar system community? How could the two choices be compared? Not unexpectedly, discussions were sharply divided. As ESA's authorised history relates: "The supporters of both missions strongly lobbied to have their pet project approved by ESA's decision-making bodies. Behind Hipparcos were the astronomers and the French delegation to ESA; support for Giotto came from the already established constituency of the ill-fated ESA/NASA cometary mission to Tempel–2, from the German delegation to ESA, and from the influential ESA Director of Science, Ernst Trendelenburg. The Science Advisory Committee also liked Giotto as it considered Hipparcos too costly, and its technical feasibility not completely established."

Two restricted sessions later and the Science Advisory Committee reached a tentative accord: it would select the Giotto comet mission as the Agency's next scientific project, but with a double proviso. Its scientific value would have to be further substantiated over the following three months. And its estimated cost ceiling of €120 million could not be exceeded. While Hipparcos had convincing advocates in the Science Advisory Committee, including Jean Kovalevsky from France and Gustav Tammann from Switzerland, British physicist Harry Elliott finessed the discussions by arguing forcefully that "a definite decision to proceed with Hipparcos could not yet be taken because of the absence of complete confidence in the technology."

Complete confidence in a space mission is quite difficult to achieve, and the comment was certainly a truism. But it scattered doubts which, once sown, were reflected in the committee's resolution. The committee requested that the three-month interregnum should also be used for further technical studies of the astrometry mission. In addition, it requested the Director of Science to find the ways and means whereby the telescope, so central to the satellite's purpose, could be funded nationally and not by ESA.

It thereby re-opened a thorny debate which had been enacted before, and which would surface again many times in the future. Should ESA simply exist to develop, finance and launch space platforms on which scientists placed their nationally-funded experiments? Or should it fund the experiments also? Managerially, technically, scientifically, politically, and financially, there are pros and cons for both points of view. The entire Pandora's Box is opened anew at the time of each major selection, those peering inside for the first time usually bemused and perplexed by the gifts lying in wait.

The committee's resolution was a potentially crippling blow for Hipparcos. For it concluded that: "In the event that the Hipparcos payload would need to be funded within the mandatory ESA programme, the committee was divided as to whether Hipparcos should then remain as the Agency's choice, or if EXUV should rather be carried out because this mission was considered by some members to be just as interesting." ESA's authorised history goes on:

> The Science Advisory Committee decision came as a bombshell in the scientific community. Klaus Pinkau [chairing the meeting] and Ernst Trendelenburg, as well as the Science Programme Committee chairman [Edoardo Amaldi] and the ESA Director General, were flooded by telexes and letters from all over Europe, blaming, on the one hand, the unusual and 'arbitrary' procedure of recommending a project not supported by technical studies and not previously discussed by the Solar System Working Group, and claiming, on the other, the great support that Hipparcos enjoyed within the scientific community. The chairman of the Hipparcos consortium, [Italian astronomer] Pier Luigi Bernacca, wrote that 170 research proposals for the astrometry mission had been presented by 125 astronomers from twelve countries, recalling that twenty four institutes from eight countries were available to put manpower into hardware and software activities, and five were already working on aspects of hardware and software using their own funds. The cometary lobby was just as active, however, and many telexes arrived expressing satisfaction with the Science Advisory Committee's decision and whole-hearted support for Giotto—which was "a once in a lifetime opportunity."

Such are the chasms of impossible decisions across which we string a thin tightrope and insist that our unsuspecting leaders traverse. And thus befell to the Science Programme Committee, at its meeting of 4–5 March 1980, the unenviable task of arbitrating on the selection of ESA's next scientific project. Big money, unique scientific opportunities, and countless careers would ride on the outcome.

Istituto Nazionale di Fisica Nucleare (INFN)

Edoardo Amaldi

The meeting was chaired by Edoardo Amaldi, Italian physicist and scientific statesman, one of the founding fathers of both CERN, the European Organisation for Nuclear Research, and ESRO, the forerunner of ESA.

There was no solution which would please everyone, and the meeting was conducted in an atmosphere which reflected this. Most delegations "regretted" the lack of information and the hurried decision that the committee was being asked to confront. The French representatives explicitly challenged the executive for presenting a proposal which was "not politically advisable since it had not met with a general concensus in the scientific community and could possibly lead to a complete split in the Committee." Only Germany and Sweden came down resolutely in favour of the Giotto comet mission. France, Belgium, The Netherlands, Denmark, Italy, Switzerland and Spain supported Hipparcos. The British delegation requested that a decision be deferred until more information was available on each, and the Irish announced an equal interest in both[31].

If a vote had been taken by the Science Programme Committee, it seems inescapable that Hipparcos would have been chosen as the Agency's next scientific project. But such a decision, Amaldi recognised, would have left several delegations, and a large fraction of the scientific community, deeply dissatisfied, for there were scientists from most countries who had some interest in each.

Instead, a compromise was negotiated: Hipparcos was reinstated as the next scientific project, with the provision that ESA should also take technical and financial responsibility for the telescope as well as for the satellite platform. The study of the Halley mission was to be extended for three further months and, if proven technically feasible within a cost envelope of eighty million euros, would also be included in the programme as a fast-track priority. As ESA's authorised history viewed it "the stars could wait, while the comet could not be stopped in its journey through the solar system, and the two-week launch window of July 1985 could not be missed."

The compromise did not make the Science Advisory Committee happy, at least not its chairman who offered his resignation to the ESA Director of Science, Ernst Trendelenburg. Facing the tight financial situation of the science programme, he did not like the disproportionately large price tag for the astronomy mission, nor the decision to finance the Hipparcos payload out of the ESA science budget. In the domain of solar system science it was customary for ESA to provide only the satellite platforms, with the experiments funded nationally and provided by scientific laboratories from the various nation states. Dual standards seemed to be operating: Hipparcos was to be the third astronomy programme, following the X-ray observatory EXOSAT and the European Faint Object Camera instrument on the Hubble Space Telescope, for which the experiment costs would be carried by ESA. The inevitable consequence would be a reduction in the funds available for new space projects.

Ernst Trendelenburg
(c. 1980)

The final decisions were taken at the Science Programme Committee meeting on 8–9 July 1980[32]. Reports of the meeting suggest that it was not at all plain sailing either, again ending in an uncomfortable compromise between ESA's two most influential member states at the time. After lengthy discussions, the German delegation agreed to withdraw its request for alternative funding for the Hipparcos instrument, nonetheless still convinced that the science budget would be blocked for a long period, to the detriment of future launch opportunities. France continued to vote against Giotto, formally on the grounds that the opportunity to provide experiments on board would not be open to scientists from the United States.

Giotto was finally adopted with ten votes for and one against. Trendelenburg stressed that "the Giotto project was certainly more risky than any other project undertaken by the Agency to date, but believed that ESA had demonstrated that it was technically able to undertake such a project, and hoped delegations would fully support the executive in its endeavours to carry out the mission successfully." Hipparcos would follow.

THE ESA AUTHORISED HISTORY sums it up perceptively after chronicling several similarly difficult choices over nearly two decades: "Choosing a big scientific project is also a matter of confrontation among scientists involved in the decision-making process: members of advisory committees or national delegations, government advisors, and policy makers. At each stage of the process, the traditional ties of cooperation, fellowship and solidarity that characterise the scientific community are strained by the emergence of national interests, disciplinary competitions, personal ambitions, career expectations, and personal relationships. When only one or two big projects can be started every three or four years the stakes are high and scientific objectivity is often a luxury. When making a choice entails some kind of painful discrimination, personal prestige, diplomatic talent, and personal or professional links can play a decisive role."

THE CONSEQUENCES OF THESE DIFFICULT but momentous decisions have echoed on down the years. With almost three decades of hindsight we can see that, ultimately, both projects were highly successful.

ESA

Halley's comet seen by Giotto

Giotto went on to make its spectacular encounter with Halley's comet on 14 March 1986, a challenging project completed in record time. Its remarkable rendezvous was broadcast live around the world, providing humanity's first close up view of a comet nucleus. The historic event put ESA in the spotlight for its stunning technical achievement, although its nascent public relations effort at the time of the fly-by was roundly criticised. Largely unscathed by its close passage, Giotto sped off triumphantly to a new encounter with comet Grigg–Skjellerup in July 1992.

The financial concerns expressed at the time were, however, surely vindicated. Following seven satellites launched by ESRO, the forerunner to ESA, in the period 1968–1972, and three successfully launched by ESA between 1975–1978, only three were launched in the 1980s: the X-ray observatory EXOSAT in 1983, Giotto in 1985, and Hipparcos in 1989. Far from being a consequence solely of the advisory committee decisions of the first six months of 1980, and attributable in part to the growing complexity of space missions, this was nevertheless a far cry from the target of one launch per year which had been suggested by the Science Advisory Committee as necessary for a viable European space science programme.

The 1980s, in practice, marked the dawn of a voracious scientific appetite. It was matched with a precocious engineering capability in Europe, driven by many exciting ideas from its member state scientists clamouring for ever more challenging and expensive missions to unravel the secrets of the Universe. A generation of brilliant space engineers were ready to meet the challenges thrown at them. And there was a corresponding industrial development, growing in technical stature and the capability to manage the resulting programmes. At the same time, this exciting period was soon to be met by a downturn in ESA's real scientific programme financial budget, and a rise in the administrative structures put in place to execute it.

But Hipparcos, at least, had its feet in the starting blocks. No ESA science mission before, and only one since, had ever been approved, only to be cancelled before launch. With its ticket to space all but guaranteed, celebrations were the order of the day.

Danish astrometrist Erik Høg, co-opted onto the Astronomy Working Group between 1976–78, and a member of the Hipparcos scientific advisory team from 1981–1997, has written almost three decades later of his conviction that, if approval had failed back in 1980, then space astrometry would never had happened. His arguments were that[33]:

> For decades up to 1980 the astrometry community was becoming ever weaker, the older generation retired, and very few young scientists entered the field. I myself would have lost faith that the astrophysicists would ever let such a mission through, and others would also have left the field of space astrometry. If someone would have tried a Hipparcos revival one or two decades later, the available astrometric competence would have been weaker, and where should the faith in astrometry have come from then? When Hipparcos became a European project in 1975, and the hopes were high for its realisation, the competence from many European countries gathered, and eventually was able to carry the mission. This could not have been repeated after a rejection.

Would NASA have picked up such a mission in subsequent years? No, believes Høg for two reasons: first, they had less breadth of competence to draw on than in Europe in this specific field. And, quoting an American colleague, "You can convince a US Congressman that it is important to find life on other planets, but not that it is important to measure a hundred thousand stars."

I MET PIERRE LACROUTE for the first time only in 1981, after the detailed satellite design phase had started. Then aged 75, he was still actively writing technical reports, developing a way to improve the measurements by making use of the dynamical stability of the slowly-spinning satellite. He made several visits from his retirement home in Dijon to ESA's technical centre in The Netherlands to discuss his ideas. He spoke a laboured English when necessary, but preferred French. Always immaculately dressed in a three-piece suit, he was uncommonly distinguished-looking amongst the broader ranks of astronomers to whom sartorial elegance is more generally anathema.

The full acceptance of Hipparcos as a space mission must have meant much to him, yet in our various meetings he showed little emotion or excitement at the impending prospects. In an obituary[34] André Heck, who had worked with him in Strasbourg, noted "I was always impressed by his kindness, although he was moulded in the old-style, somewhat authoritative, managerial approach."

Over the next decade, Lacroute watched calmly from his home in Burgundy as Hipparcos slowly took shape, joining us for the launch from Kourou in 1989.

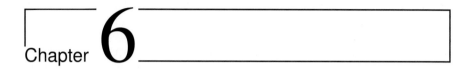

Chapter **6**

From Concept to Launch

Projects prosper if there is a powerful, centralised, unified project management team in place, with a project manager who is responsible for the project from cradle to grave. Once this manager has become familiar with the proposed task, the first essential job is to specify the resources of money, staff, time, and facilities required for completion and to offer milestones of achievement along the way. The whole undertaking is likely to take many years, during which period none of the key staff should change. If the task is successfully accomplished in the time and with the resources they specified, a double promotion should be the reward; if they fail to deliver, retirement may well be appropriate. By contrast, insufficient authority for the management team, with frequent changes of its personnel, is a sure recipe for disaster.

Professor Sir Hermann Bondi, letter to *The Times*, 31 May 1995

I HAD JOINED ESA IN JANUARY 1980. I had a degree in theoretical physics, and arrived with a PhD from the Cavendish Laboratory in Cambridge. My research had been in Nobel Laureate Sir Martin Ryle's radio astronomy group, where I had been investigating the nature of distant radio-emitting galaxies.

Hipparcos was unknown to me, and I had followed neither its concept studies of the late 1970s, nor the political struggles surrounding its eventual adoption which were being played out around me during my first months in The Netherlands.

I knew little enough about stars, and could barely have imagined spending the following twenty five years of my life in the field. But ESA was looking for someone to take on the role of coordinating scientist for Hipparcos, a field new to the organisation also. Brian Fitton, head of the astronomy division at the time, suggested that I take a look at the project, to see if it held any interest.

Although measuring star positions had sounded uninteresting to me, I had not appreciated how ingenious the new technique was. Neither had I been aware of the advances in science which could be expected as a consequence of its successful execution.

Reading the description drawn up as part of the feasibility study, and on which the various advisory groups had based their own judgments months before, I was immediately captivated. The satellite concept was delightfully elegant in its observing principles, and it was a masterpiece of instrumental creativity. I will try to describe these features in outline. Equally remarkable were the mathematical manipulations necessary to construct the resulting star catalogues on ground, which I shall not.

THE WAY THAT THE MEASUREMENTS were made can best be explained in three conceptual steps. To start, imagine looking up at the night sky in a Universe in which everything is at rest. You are on an Earth which is neither rotating nor moving in orbit around the Sun, the Sun is not moving through space, and the distant stars are fixed points of light. In this simplified picture, the location of each star is described by just two coordinates, corresponding to angles of latitude and longitude on a map of the Earth or sky.

Conceptual step 1, with stars fixed in space

The figure shows three such stars from one area of the sky a few degrees in size (filled), superimposed on three stars from another field some way distant across the sky (open) because of the split-view mirror. Now picture a swathe of a degree or so in width which covers a full circle around the celestial sphere, crossing some of these stars. And imagine measuring the angles connecting successive pairs of stars as we step our way around the sky along this swathe. Back to the starting point, these pairs of angles would add up to three hundred and sixty degrees. Our measurements would give an estimate of the star positions stepped around this circle, one with respect to another.

A second circle laid out across the sky in a different orientation gives another set of angular measures between all the stars in that new circle. With lots of these circles, at all sorts of different directions crossing the sky, the result would be a dense network of many different pairs of angles between any chosen star on the sky, and the many others all around it. The satellite from its vantage point high above the Earth would turn slowly across the heavens, completing one measurement circle in just over two hours. In its three years of observations it would trace out ten thousand of these circles. By continually twisting its spin axis as it scanned, the circles criss-crossed the sky in all sorts of directions. In the process, each of the hundred thousand catalogue stars was observed more than a hundred times, with each star being connected to countless others.

With the thousands upon thousands of measurement arcs, and with the linking up all of these triangles, what would drop out of the computer reconstruction on the ground is a map of the stars with the position of each one relative to every other. The resulting map is very rigid; every star is snapped tightly into its place by the dense network of other measurements. Three years later—only a few weeks or months of observations are not enough—and a full map of positions across the sky would result. No time is wasted in the survey: the satellite just spins round and round very slowly, building up its network of measurements for analysis on the ground.

The rigidity of the map is greatly strengthened by the two widely-separated viewing directions. Imagine a 4-armed windmill at night with its arms turning slowly, a telescope attached to just two adjacent arms, somehow able to see the whole sky as if through an invisible Earth. Labeling positions around a circle as if laid out on a clock face, then stars at 12 o'clock are 'connected' to those at 3 o'clock. A little later on, as the arms turn, the stars at 3 o'clock are connected to those at 6 o'clock. Similarly those at 6 are connected to those at 9, and later those at 9 to those at 12 again. Sadly, those at 1 o'clock are only ever tied to those at 4 o'clock, those at 4 to those at 7, and so on. So we would have a kind of rigidity, but the pieces of the circle aren't themselves very well connected together. The situation improves almost magically if the angle between our two telescopes doesn't divide exactly into a full circle—Hipparcos chose fifty eight degrees. Then, as the structure turns, and after just a few rotations, all parts of the sky have been connected up, and the star positions are pinned down very tightly indeed.

THE DESCRIPTION OF THE METHOD started by imagining that the stars are fixed in the sky. Proceeding one step further towards reality, now assume that each star is no longer stationary, but instead is moving on a straight path through space. It's important to recall here that their distances are so vast that they move only through very tiny angles, even over a year or more. They can be moving towards us, away from us, or in any other direction, and with any velocity. Our angular measurements always allow us to pick out the star's proper motion, that part of the star's angular movement projected onto the celestial sphere.

This means that although all the stars are moving, in a seemingly random way on the sky, we only need two extra numbers to describe their motion. So we now have four numbers describing each star: one pair tell us its coordinates at a chosen moment in time, and a second pair tell us how it moves across the sky over one year.

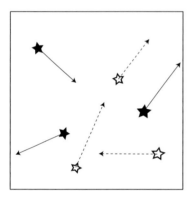

Conceptual step 2, with stars moving through space

Even though the relative positions of all the stars appear to be changing in a very complicated way with time, the mathematical description of this motion is actually very simple. Just these four numbers are all that's needed to describe each star's resulting position, even over a period of several years. Again, the fact that we have made a hundred or more observations of each star over a time interval of several years, means that these four numbers are extremely well nailed down.

At this point, I stress again that we have not measured the true space velocity of each star— just its angular motion on the sky. As for the aircraft analogy earlier on, we would need to know the distance to convert the apparent angular measure into a true space velocity.

FOR THE THIRD STEP IN THIS EXPLANATION, we now need to include the fact that each star has its own, unknown distance from the Sun—some are far away, some are relatively nearby. As we have seen, each star's distance can be described by just one number, its parallax. This just provides a direct measurement of the star's distance through a knowledge of the Earth's orbital dimensions. The key point for the measurements is simply that each star's back-and-forth motion during one year is described by just one single additional number, the size of this parallax angle.

The final picture is in fact very simple: although the stars will be moving one with respect to another in a very complex way depending on their space motions and their distances, the motion of each star on the night sky is fully described by just five numbers: the two position coordinates, the two proper motion coordinates, and the parallax angle.

Together these five numbers describe where the star is at a particular moment in time, how it moves through space, and how far away it is. Astronomers currently use the start of the year 2000 for the reference time, and the star positions in the published catalogues are adjusted to this specific moment. If we need to know where a certain star is on, say, 21 September 2014, because we want to point our telescope there on that date, we take the catalogue position at the start of 2000, and add just over fourteen years of the star's proper motion through space.

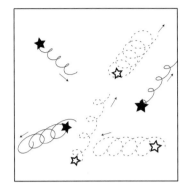

Conceptual step 3, with parallax included

From this celestial jigsaw of measurements, from this interminable sequence of angular separations, stepping around the sky over three years, we can reconstruct their motions and their distances; their stereoscopic, three-dimensional distribution, and their motion through space. There's nothing particularly special about the three years of measurements either. Longer would have given results a little better, but there is always a balance between the quality of the results, the cost of operating the satellite, and the time taken to do the computations on ground. There are, of course, endless scientific and technical complications which in reality accompany this simplified picture. However, none affect the basic principles.

If the preceding description is a little confusing, the essence is simple: all stars are moving through space, and all appear to move through the effects of parallax as the Earth goes round the Sun. Measure their relative separations carefully and repeatedly, and we can reconstruct all of these movements.

At the time of its acceptance by ESA in 1980, the project was, it must be admitted, little more than a concept on paper. There was a description of the science motivation and measurement principles, a rather detailed explanation of how the observations could be carried out in practice, and an outline of the number-crunching needed on the ground. After the battle for selection, a new challenge began: to come up with a detailed instrument design and, after that, to build the satellite itself.

AFTER HIPPARCOS HAD BEEN accepted by ESA, the first high-level task inside the organisation was to select a project manager, and a project team with him, who would take responsibility for the project, and report up through the management line of the scientific directorate. Ernst Trendelenburg, ESA's director of science at the time, duly nominated Italian engineer Franco Emiliani to fill the role.

Franco Emiliani (2007)

Emiliani had come from an early career in the Italian navy working on surface-to-air missiles. In ELDO from 1967 overseeing construction of the launch facilities in French Guiana, and thereafter in the Spacelab team in ESA, he trod a balanced line between forceful leadership and a courteous and engaging style. He duly put together a team of twenty or so engineers who would assist him running the project from the ESA side. Their task would be to prepare and manage the industrial contracts, work with Arianespace to prepare for its launch, liaise with the team at ESOC who would operate the satellite once in orbit, and keep track of costs and schedule.

The project team was divided into four sections: one responsible for the experimental payload, one for the supporting spacecraft systems, one in charge of the assembly and verification, and one for the overall performance. Over the next eight years, the team directed the progress and monitored the costs of the large industrial teams delegated to carry out the detailed design and subsequent construction.

In addition to his technical and management skills, Emiliani had a significant curiosity in the science underlying the mission, a keen interest in understanding the details of how it worked, and considerable respect for the scientists who were driving the project from outside ESA. They are not universal characteristics of managers of large projects, for whom the actual goals may be of only passing interest. But an enthusiasm for the underlying purpose is an enormous asset, and in this case contributed to the motivation of the entire team, facilitated the technical negotiations with industry, and contributed to the makings of a collaboration between the space agency and the scientific community which worked particularly well.

Another most important straightjacket was imposed on the team. When the project had been accepted by the Science Programme Committee in 1980, it was on the basis of the accuracy that the earlier study had indicated would be possible—two thousandths of one second of arc for the positions and parallaxes of every star. Emiliani accepted the challenge, and passed it on to his team as their overarching objective, their Holy Grail that would be ESA's, industry's, and the science community's demanding quest and enduring mantra for the long decade ahead.

In such a project, ESA is the customer. Through the project manager, the customer sets the specifications, the budget, and the schedule. Once contracts are awarded to the industrial teams capable of designing and building the satellite, the project team agrees (or otherwise) to the solutions proposed, tracks the performances, costs, and schedule, and arbitrates on any

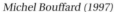

Michel Bouffard (1997)

Bruno Strim (1997)

disputes. For those tempted to think that the task might be a light one, how many have engaged a contractor for house repairs, and duly overrun their starting budget?

The main industrial contracts were awarded, after keen bidding, to two major industrial teams. To secure the contracts—which would guarantee business, jobs, and prestige for years—their bids had to be extensive and meticulous, describing their company's experience and the individuals that would be assigned, their preliminary designs, technical issues, and detailed planning of schedule and cost. Each bid from the two leading contractors comprised some twenty ring binders of technical minutiae, a couple of meters shelf space of documents, each of which had to be reviewed with care and attention. Matra Marconi Space[35] in Toulouse in France, under their manager Michel Bouffard, was duly awarded the prime contract for the telescope, the detection system, and all the associated experiment parts. Alenia Spazio in Torino in Italy, under Bruno Strim, would be responsible for the rest of the satellite support systems, and the overall satellite integration.

FROM ACROSS THE FAR-FLUNG nations of Europe, industrial teams with specific knowledge or distinct expertise bid, in their turn, for a lower-level task. Hundreds of contracts were signed off. In this way, around thirty different industrial teams from many countries pooled their expertise, and would become involved with the design and manufacture of the various components. Across the various industries, the workforce dedicated to one piece of work or another probably numbered around two thousand. Together, and orchestrated by the prime contractors, they then came up with drawings and descriptions of the overall system and its various subsystems—the power supply, the attitude control, the thermal control, the telecommunications, the telescope optics, the on-board processing, and so on. Manufacturing procedures were elaborated. Protracted negotiations on costing, construction, and testing edged forward.

ESA's Hipparcos project team (1985)

The challenge of running a project of such complexity is firstly to find and recruit the army of workers with all the skills required. But even more crucially, to ensure thereafter that all of the many and varied interfaces fit together, and that the whole effort is organised to come together on the schedule demanded. The launch contract, once signed, is rather immovable and, in any case, delays are expensive once a large team is in place.

Over the next three years, the satellite progressed from a conceptual design on paper, to a detailed description of the overall system and its component parts. The system section built a detailed model of the predicted performances. This meant that we could see, at a glance, what would be the effect on the final accuracies of changing the mirror size, its reflective coatings, the spin rate, the amount of stray light falling into the telescope, and a whole host of other details.

Hundreds of meetings, thousands of documents, and countless reviews of the component parts were clocked up as the weeks and months passed. Major reviews were held to look over the entire system design. Contract and cost meetings between ESA and the various contractors paralleled the technical and performance meetings. All of the major steps followed the detailed schedule drawn up at the start—for the costs to be held in check, the schedule was paramount.

Eventually, around 1984, the blueprints were finalised. The project had moved from concept to detailed design. To mark such a major milestone, it was a tradition for the leading contractors to host a small party for the customer. Toulouse and Torino were, naturally, excellent cities for culinary celebration, and we could enjoy rare moments of more relaxed discussions outside of the meeting rooms.

THE DESIGN AGREED, it was then time for the various industries to start on the construction of the individual parts. Some components would be fairly routine. Others would represent challenges at the very limits of technical feasibility, and would sit on the critical path, thorns in our side, for the following three years. Once the parts were manufactured, they would be integrated to form the complete satellite. Detailed testing would take place along the way.

Satellite construction usually proceeds through a series of prototypes to test the design: in this case, Emiliani had agreed with industry that they would provide an optical model to verify the quality of the telescope and its alignment before final manufacture, a thermal model to be tested in the conditions of solar illumination in space, and a mechanical model to be shaken to simulate the launch conditions. Once these different parts were validated, the flight model would be built, assembled and tested in its turn.

Industrial progress meeting with Alenia (1986)

Franco Emiliani moved on from the Hipparcos project in 1986, his place as project leader taken by his spacecraft section head, Hamid Hassan, who thereafter held the post until launch. Hassan was a different character, whose distinct management style was equally well matched to this second phase of the development. Here, design creativity was no longer relevant and no longer sought. Schedules and deadlines, industrial contract obligations, instrument performances, and financial control were instead paramount. He built good relationships with the industry prime contractors, and he and his team got the very best out of them as a result.

Hassan also had great esteem for the scientists working on the project, and the two of us—he as project manager and me as project scientist—had an excellent relationship founded on trust and a respect for the very different, but equally indispensable jobs that we both had to exercise. Both in the same boat, we had to steer a common course. At weekly project meetings we would identify the challenges ahead, and agree on any compromises.

Hassan's wry sense of good humour and self-deprecating demeanour masked a sharp intelligence, and he would occasionally goad me with a favoured quip, each time telling me as if it were something witty and original which he had just thought up: "We project managers don't mind if we're told to launch a ball of wax on a piece of string—our job is to get it launched on schedule and within budget."

In reality, Hassan was immensely proud of the task he'd been given, the team of individuals that he had been asked to lead, and the place in history that he knew Hipparcos would later occupy. The satellite was duly launched on schedule and within budget, all technical challenges resolved along the way, probably one of the very few comparably-sized projects which could make that claim.

The words at the start of this chapter were from the mathematician and cosmologist Sir Hermann Bondi. Bondi had been Director General of ESRO (the forerunner of ESA) between 1967–1971, and thereafter the UK's Chief Scientific Adviser to the Ministry of Defence between 1971–77, Chief Scientific Adviser to the Department of Energy from 1977–1980, and Chairman of the Natural Environment Research Council from 1980–1984. In an earlier life he had even written learned articles on magic squares. With this weighty portfolio, he might not have been able to claim that he had seen it all, but he had probably seen enough. His comments targeted the massive cost and time overruns in recent Ministry of Defense projects that had been reported by the National Audit Office.

Bondi articulated that both cause and cure for such overruns were well known, and yet the solution often raised insurmountable difficulties for career structures which all-too-often presuppose frequent changes of post. In short, Bondi was critical of projects and organisations with changes of top-level managers engaged solely in career hops, moving on long before accountability caught up. His principles were broadly followed for the development of Hipparcos, and we can look back on a decade of cutting-edge space engineering and say that it worked. It is regrettable that these simple and somewhat self-evident prescripts are not always adopted.

BACK TO 1980, A PARALLEL ARM of ESA's science directorate supplied the mission's scientific leader. In this capacity, my task was to ensure that the instrument worked as it had been laid down in concept, and to optimise the science that could be done within the budget assigned. To ensure that the scientific objectives remained paramount, I was not accountable to the project manager. My influence on the project was through scientific persuasion rather than fiscal authority.

In these aspects I worked closely with Maurice Schuyer's system group within the ESA project team. But I also had the benefit of one extensive resource at my disposal, if only it could be coordinated and properly channelled: the European scientists and astronomers interested in the science, pushing the mission since its earliest awakenings, ready for deeper involvement in all of its aspects, and committed to work to its full success. My task would be to marshall and coordinate their efforts, use their knowledge to guide the design, manufacture and testing of the satellite on ground, and to advise and assist with its operation once in orbit.

Maurice Schuyer (2007)

One example is sufficient to illustrate the process. The groups preparing the analysis of the data knew that they needed a knowledge of the spacecraft velocity at all moments to correct for the aberration of starlight. Somebody would estimate how accurately this would be needed, and write a technical note giving their reasoning. Others would review the results. The resulting specification was passed to industry to make sure that the spin control could deliver the required performance, while the operations centre would propose a methodology for measuring it. At any one moment, hundreds of these scientific threads would be spinning their way across Europe, backwards and forwards between the relevant players to ensure convergence.

Something of a complication was that, in all of this scientific effort, no funding from ESA was available: each scientist or scientific team would be wholly dependent on their own funding, typically through university or national grants. They would have to petition for additional research positions, computer resources, travel funds, and whatever else they needed to do their work, and collaborate with others where funding was inadequate.

It may seem surprising that such a system can possibly work. How could an organisation coordinate the work of a large number of scientists, with no payment changing hands? It is a common feature of the ESA missions, and works for the following reason. Individual scientists with a commitment to the subject, would see Hipparcos as a once-in-a-lifetime opportunity for their particular field. A group with a track record in Galactic structure studies in France, an individual with a particular interest in stellar evolution models in Italy, a collaboration committed to improving the stellar reference frame in Germany, and countless others, understood that the project now starting up would provide frontline research opportunities for them in the years to come. They would lobby their own funding authorities, persuade their university astronomy or computing departments, and push their national technology institutes to get involved.

Funding might take months or years to fall into place, but the resulting resources mobilised can be enormous. The entire project being vastly more than any individual research group, or even nation, could begin to contemplate, collaboration with each other, and notably with ESA made enormous sense. In its turn, the credibility of affiliation with an international space mission would furthermore help their national funding requests. Indeed, delegates to ESA's advisory bodies all the way up to the Science Programme Committee were often senior figures within the decision-making structures in their own countries. Once a mission was underway in ESA, senior policy makers sympathetic to the cause would generally be keen to support further involvement by scientists in their respective countries. It was a fast way to gain further access to major cutting-edge research facilities.

Like a huge snowball gathering both mass and momentum, scientific and technical interest grew in the member states. A key problem for me would be to keep all of this manpower prioritised, focused, on track, and especially given the unpredictable timelines so frequently encountered in research efforts, on schedule.

MY FIRST TASK AS THE PROJECT got into full swing in 1981 was to set up a scientific advisory group, the Hipparcos science team. This would be composed of scientists from universities around Europe who would together represent all of the many scientific disciplines needed to design the instrument, and make sure that no effect had been forgotten, no complication omitted or left to chance.

The *modus operandi* for such science teams had been established by other projects over the years. Once in place, it came together for a two-day meeting three or four times a year, under my chairmanship, to run through all of the scientific and technical issues which needed discussion, resolution, and agreement for the project as a whole to advance. Relevant engineers from the project team were also present. Tasks that needed to be investigated further were assigned to a relevant individual: perhaps someone from the project team who might in turn pass it as a task to industry, or one of the science team members best placed to work on the problem back in their home institute. To keep the project on track, a schedule for each task was agreed, and results were reviewed at the next meeting.

It is, I believe, of paramount importance that one person knows approximately what is going on in every part of a project of this size, maintaining a top-level picture of how all the puzzle pieces should fit together. This was one of my tasks. Like a conductor in a large orchestra, I couldn't play any of the instruments particularly well myself, and I had neither the aptitude nor time to learn. But I would have to maintain the synchronisation and set the tempo through to the final bar. Only then might the audience applaud.

Two related problems stood in my way in setting up the science advisory team. The first was that I had no background in the field and therefore no knowledge of the leading scientific exponents, nor of their specific skills and temperaments. Second, of which I was conscious at the time and perhaps more so in retrospect, was my age. At twenty six, I could expect a rocky road ahead in gaining the confidence of the Europe's astrometric leaders. In both I landed on my feet, and for similarly related reasons. The leading astrometrists around which the studies had been structured were small in number, and as new to the opportunities of space as I was.

At my first meeting with the leading players, in Paris, convened by the experienced and unflappable secretary of ESA's Astronomy Working Group, Dutchman Henk Olthof, I was greeted as a collaborator and, in showing my own commitment to the project, soon as a colleague. Catherine Turon, from the spectacularly situated Meudon campus of the Paris Observatory, Lennart Lindegren from Lund University in Sweden, Erik Høg from the Copenhagen University Observatory in Denmark, and Jean Kovalevsky from the CERGA institute perched in the hills of Provence above Grasse in France, were all hugely courteous and motivated only in moving the Hipparcos project forward. They welcomed me as the missing link of some grand overall plan, and we got down to work.

Henk Olthof (2002)

The advisory team became the scientific voice of the project in Europe, and it was through their guidance that the scientific aspects—longer-term strategic problems as well as more day-to-day issues—were monitored, steered, anticipated and resolved. The team would have to guide the project over the years to come, and it was crucial that I got the composition right. Expertise was needed to cover all of the major sub-fields of the project: optics, detectors, instrument design, calibration in orbit, astrometric needs, existing catalogues, and data analysis. Such a team should not be too unwieldy in size—perhaps a dozen people at most. It should be selected according to competence alone, although as a pan-European project, broad geographical representation would be advantageous. The members had to be used to working to agreed schedules—in a team with many players, working in a project with many complexities, and comprised of scientists with other duties, this wouldn't be easy. It had to be a team that would work well together, and with the common aims and ultimate scientific objectives firmly in mind.

The choice, in reality, proved not too difficult: the scientists leading the studies carried out so far were relatively few in number, worked well together, and by-and-large covered most of the disciplines that would be needed. I talked to each, synthesised the totality of their own thinking, and assembled the guiding team. The first meeting was held at ESA's technology centre in The Netherlands on 28–29 April 1981. It remained in place for seventeen years. The thirty-ninth and last meeting was held on 9–10 October 1996, when it gave the go-ahead for the final catalogue release.

I met Sir Hermann Bondi around the time of the catalogue publication. I should have mentioned this continuity of effort to him. He might have been interested. I'm sure he would have approved.

THE TELESCOPE WAS CENTRAL TO HIPPARCOS, and I brought in an independent optical expert, Charles Wynne, at the very start. Then aged seventy, he was a significant figure in the design of optical instruments, and in that capacity had received the gold medal of the UK Royal Astronomical Society in 1979, its highest award. He was known for what was almost a monopoly of original designs for wide-field optics for large telescopes, for a series of scientifically elegant spectrographs, and neat systems used to correct the effects of atmospheric dispersion. He had had nothing to do with the Hipparcos studies to date, which left him well placed to examine dispassionately the optical designs put forward by industry.

René Bonnefoy (2007)

The telescope design and construction in reality represented the combined efforts of many different groups. René Bonnefoy in ESA led the part of the project team focused on the scientific instrument, and he had overall responsibility for making sure that the telescope design was plausible, that it could be built, polished and calibrated by the industrial teams, and eventually launched and operated in space.

But getting the very basic design of the telescope correct was a tremendous challenge. There are many different telescope concepts—Newtonian, Cassegrain, and Ritchey–Chrétien amongst them. Named after their eminent designers, they differ in the details of how their various reflecting surfaces are polished and arranged. They nonetheless share a common objective: to create an image at their focus. The skill of the telescope designer's art is to consider what properties of the particular system are most critical—whether it be the largest field of vision, the smallest image distortion, the least colour imperfections, the highest throughput, the broadest wavelength response—and to design the system required.

Very often, as for Hipparcos, all of these are important, and it falls to the scientific and engineering teams to weigh up the various pros and cons before agreeing to a particular design. Whether it can be built, polished, and aligned, are additional practical considerations. Hipparcos needed to be able to look out in two widely separated directions at the same time, and to bring the images from the two fields to a common focus. Such a telescope had not been conceived of before, let alone built and deployed in space.

Many possible options were investigated, and their strengths and weaknesses circulated in technical documents for analysis and discussion. The solution finally proposed by the industrial leaders was daring. They would first polish a single mirror to a special shape, carrying the profile of a

Schmidt-type telescope specified by a detailed computer ray-tracing analysis. Just thirty centimeters in diameter, this main reflector at the very centre of the satellite would not be much larger than a decent-sized shaving mirror. They would then slice it in half without disrupting the surface quality, rotate the two halves by an angle demanded by the two viewing directions, and then glue the two pieces back together at this chosen angle. Other mirrors in the telescope would have their own special surface shapes, all optimised by the computer design.

Once assembled, the two incoming light beams would bounce off the split mirror, be intercepted by the mirror next in the telescope's path, and then projected on to a common focus where more magic would be woven.

So simply described, the task was forbidding in practice. For a start the design demanded an assiduous polishing, accurate across its surface to a fraction of the wavelength of light. If we imagine inflating this modest-sized mirror to the dimensions of the Atlantic Ocean, the residual lumps and bumps of the scaled-up monolith could not deviate from the underlying smoothness by more than ten centimeters in height. The mirror slicing could not be allowed to affect this delicate surface either. And when it came to glueing the two parts together, it had to be with meticulous precision, and using an adhesive that would hold the two parts rigidly in place during the launch, and yet not deform the bonded surfaces as the satellite aged or 'outgassed' once consigned to the almost perfect vacuum of space.

It was an even more complex approach to this mirror at which even the enormously competent French Space Agency CNES had baulked during its own studies a decade before. And it would not be the first time that a wonderful telescope designed on paper might not be possible to fabricate in practice. The famous Cassegrain reflector, developed by Laurent Cassegrain in 1672, had been invented independently at least three times before that, including by Marin Mersenne, the 'father of acoustics' who also gave his name to an important family of prime numbers. All were prevented in their attempts to create the actual telescope by the available technology, Cassegrain included. Our success in transposing paper design to glass could not yet be guaranteed either.

Pierre Hollier (1992)

When the industrial teams started the satellite design, it was their creative optical expert Pierre Hollier from Matra Marconi Space in Toulouse, graduate of the Ecole Supérieure d'Optique in Orsay, who stepped in with a feasible telescope concept, and Lennart Lindegren from Lund Observatory in Sweden who pointed out a particular trick which allowed the dissection and reassembly to work. The telescope images, Lindegren pointed out, could be improved by carving off and discarding a sliver from the two halves before re-bonding. I have heard it said that nobody is quite sure how Bernhard Schmidt came up with his practical method of making the difficult corrector plate in the revolutionary wide-field telescopes that he invented in the 1930s and which now bear his name. Sixty years later and I'm not sure any of us understood from which hat Lennart Lindegren plucked his slicing-and-dicing magic, but it was just one of a number of insightful and indispensable legacies that he brought to the mission. It typified the sort of scientific creativity that few can emulate, and yet which decides the fate of these complex systems. We could only read his technical note on the subject with admiration, and shake our heads with bemusement.

I have the very greatest regard for the skills of the high-technology industries who worked with us on Hipparcos. But they can be an optimistic breed, eager to get a concept pushed through, with the customer left to pick up the bill as mounting practical problems have to be ironed out later on. It can be useful to ask experts to stand back from their detailed involvement and give a top-level feeling of how things are going. Experience and intuition based on a lifetime's work are important. Early in the design phase, during one of the advisory meetings in Noordwijk, my insight at a loss to see whether the proposed mirror would serve its exacting purpose, I pushed Charles Wynne for an answer to a simple question "In your opinion, will Matra's telescope design work, yes or no?" He had clearly been thinking similar thoughts because his answer was out before my question was fully formed. "I'm not saying that it won't", he said, eyes twinkling mischievously as he hedged, "but I do know that seconds of arc don't split into milliseconds of arc very easily!"

The 30 cm diameter beam combining mirror

It wasn't the answer I'd hoped for, but it was wise council nonetheless. Three further years of effort vindicated the design; it did work, and the telescope would eventually take up its sentinel position in space, peering out, skew-eyed, across the expanses of our Galaxy to scrutinise the stars.

Once designed on paper, by computer ray tracing, this complex mirror was polished by craftsmen of the French industrial company REOSC Optique, a group specialised in high-precision optics for science and industry. Based in

Saint Pierre du Perray, south of Paris, REOSC's eventual hard-won success with this difficult mirror augmented the company's capabilities and consolidated its growing reputation. REOSC went on to polish some of the greatest mirrors in the astronomical arsenal at the turn of the second millennium, including the four gargantuan eight meter diameter mirrors of the European Southern Observatory's Very Large Telescope, now operating in concert on the sawn-off mountain top of Cerro Paranal, deep in the Atacama desert in Chile.

The split mirror was just one of a number of exquisitely flawless mirrors inside the Hipparcos telescope. The others, more classical spherical mirrors, or perfectly flat 'folding' mirrors inserted to keep the overall telescope size down to its most compact, and drilled from the back with cylindrical holes to minimise weight, were built by the German optical masters, Carl Zeiss GmbH at Oberkochen. Once polished and assembled, this unusual split view telescope brought star light from the two viewing directions to a common focus, precisely as had been commanded.

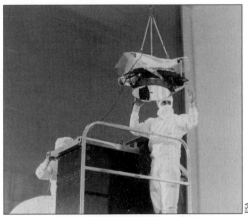

Spherical mirror integration (1987)

The optical parts were assembled to form the complete telescope by the experienced integration team from Matra Marconi Space. Polishing each surface had been a tall order. Alignment of all the surfaces to the sub-micron prescription of the computer design posed its own tricky challenge. Holding the mirrors precisely and rigidly in their designated positions thereafter was a whole specialised craft, and needed its own ingenious solution. Computer analysis showed that three support struts would be needed to hold the spherical mirror in its place during the hefty vibrations of launch. But once in space, the reduced gravity would change the forces on the mirror, and the three clamping points would cause the mirror to buckle. Another headache, another imponderable, another unexpected obstruction thrown in our path. The solution—there often is one if you assemble the right experts—was a small pyrotechnic device which would be detonated once in orbit, to cut the anchoring point of one of the struts, reducing the clamps to just the two.

At each place in the satellite where a clever design on paper had to be translated into practice, these kinds of complications invariably surfaced.

The relationship between project scientist and project manager is always a delicate one for these reasons. My brief as chief scientist was to make sure we had the clever design in place that thought of everything. Only by stepping through each consequence of every scenario would potential problems be anticipated. Emiliani's and later Hassan's brief as project manager was to translate these lofty ideals into substantive form, and to inspire and coordinate industry to deliver a working satellite, *sui generis.*

Only by bringing together the appropriate teams of creative and specialised experts could such a one-off experiment be designed and built. Yet always hanging over us was the perennial problem for a space mission: though fearsomely complex, the entire system could never actually be tested at the levels it would need to perform at in space—under zero gravity, in total vacuum, bathed in harsh ultraviolet light, and bombarded by high-energy particle radiation. Each problem was tackled rigorously, but we also kept our fingers crossed for several years nonetheless.

Much later, two years before launch, test engineers at the ESA centre in Noordwijk were examining the quality of the entire telescope before its final integration into the satellite. Designing and building the optics was far from easy, but setting up the testing equipment to demonstrate that performance was also a challenge. The telescope was cosseted in a 'clean room' to exclude even microscopic dust particles from contaminating the optics, and supported on massive granite blocks to minimise effects of external vibrations, however miniscule.

As the battery of tests proceeded, the team in charge were perplexed to find small, regular but quite unexpected motions of the star images being projected through the entire telescope. Like surgeons involved in a complex operation, the team in white coveralls, face masks and hairnets, consulted their battery of diagnostic machinery with consternation, eliminating the possible causes one by one. The source of the pulsating image motions was eventually traced to the regular impact of ocean waves pounding the beaches of the North Sea, seismic shocks felt more than a kilometer away across the coastal dunes of Holland.

Again, I could not help thinking about how this sensitive instrument would take to being launched. Had the vibration experts done their parts of the design correctly? Had the relevant launch team measured and communicated the spectrum of launch vibrations correctly? Damping vibrations is a crucial art in space projects, where resonant frequencies can wreak the type of havoc seen in the Tacoma Narrows bridge collapse of 1940. Were there other malevolent celestial equivalents of the pounding North Sea breakers that we'd not thought of, and which were lying in wait for us, high above?

Our science team meetings, usually held in ESA's research and technology centre in The Netherlands, were always an opportunity to brainstorm. It's an irritating word, but it nicely conveys what was a significant and crucial part of our collective scientific work. I would prepare the agenda for each meeting a couple of weeks in advance (in those days, circulated by telex) based on the main problems being faced or anticipated, and where scientific advice or compromises were debated and agreed.

ESA technology centre (ESTEC), Noordwijk

Members would come with the results of their own work, and perhaps recommendations or solutions, and usually with their own problems to lay on the table. Difficult issues would be lobbed back and forth until a possible solution could be sensed, and a way forward agreed. Project team members, operational team representatives from Germany, and industry experts from around Europe might attend for specific issues. Coffee breaks were times for more uncertain thoughts to be aired, or for one-on-one discussions to plan a way forward on some particular topic that didn't need to be discussed by all. A social dinner was mandatory at the end of the first day of the two day meetings: a respite from the difficulties of the day, more relaxed deliberation of the less technical problems, and an opportunity to inspire each other with thoughts of the next major milestone ahead.

IF THE BASIC TASK OF THE INSTRUMENT was to project the star images onto the telescope focal surface, the rest of the satellite had to be in place to supply it with its variety of support needs. We needed electrical power from the solar arrays and chargeable batteries, and demanded scrupulous shielding from scattered light from the Sun, Earth and Moon. High on the critical list was the ultra-rigid structure to hold all of its optical parts in sub-micron level alignment. All components had to be clamped down to prevent vibration, isolated from the effects of the launch and the boost motor firing, and stabilised against any thermal variations which might throw its highly-delicate measurement path out of alignment.

Three solar arrays would give power for the satellite to operate in orbit, its electricity generated by incident sunlight. All three panels would be folded flat against walls of the hexagonal satellite body, hinged and clamped for launch, to be opened out once in orbit by ground command. This panel opening would be another tense moment, scheduled for several days after launch. Deployment failure of one or more panels would be another death sentence, for not only would the power supply be curtailed, but the smooth spinning motion of the satellite would be lost.

On-board computers were needed to calculate which stars were to be observed as the satellite rotated slowly in space. Swift and accurate calculations had to be done in a flash to pilot the sensitive detectors to the chosen star images, and then to queue the data acquired for transmission to the ground. Specially designed antennae were needed for this too. As the satellite turned in space, different parts would face the Earth, and two antennae were needed for full coverage. The data had to be switched between the two antennae depending on which of the pair were on the sight-line to the ground. All these calculations had to be done autonomously onboard the satellite, for there was insufficient time to send data to the ground, perform some calculations, and send the results back up. The round-trip travel time for the radio waves to propagate would alone take a quarter of a second, and that was too long.

All of these systems were complex and intricate, usually demanding a specific development and manufacturing effort within the relevant industrial groups. Most components that are used in space rely on decades of space engineering experience and heritage, in materials, integration procedures, and testing methods. Items that had been 'space-qualified' in the past gave no particular cause for concern. Yet the reason that industries relish space experiments is to extend their own expertise by doing something new, gaining new knowledge, and extending their own competitive advantage. This is why cutting-edge science and high-tech industries make such a rewarding partnership. It's also one of the reasons why space experiments are so risky.

ONE OF THE SUPPORTING TASKS of the satellite was particularly challenging, and absolutely crucial to the mission's success: how would we spin the satellite to sweep out the circles on the sky? The whole satellite had to rotate slowly but precisely, following an accurate pre-defined and endless loop around the celestial sphere, on and on for three years. Several challenges had to be mastered: once set in its delicate slowly spinning motion, small adjustments would be needed to keep the two directions of view on their pre-defined track. These adjustments would need to be carried out most gently, without transmitting unwanted vibrations through to the rest of the measurement system. The designers needed to figure out how to follow where the satellite was pointing, and needed to find a way of adjusting the motion so that it could continue to point to where it needed to be over the coming seconds and minutes of its sky scanning.

The solution had the feel of something precarious if not preposterous. We would know the approximate positions of the stars that we would be measuring. We needed a way of recognising the star field that the satellite was pointing to, and thereafter making adjustments to its scanning to point it where it needed to go next. Implementing this required a plethora of sensors, monitors, and methods for the attitude adjustment, which would work together to get the job done. Once the satellite had been placed into its final on-orbit location, a coarse sensor would detect the light from the Sun, and allow the solar arrays to be pointed roughly in that direction. The satellite would then be spun-up to its required spinning rate. This motion would be sensed by gyroscopes, and adjusted to give precisely the desired spin. Next, the entire satellite would be slewed so that its spinning axis was offset by the chosen forty three degrees from a line to the Sun.

Thus gently rotating, star images would pass steadily across a specially designed star sensor at the telescope's focus. This star tracker would sense the passage of stars across the focal plane in both coordinates on the sky. The signals would be analysed at once by the computer on-board. The computer would interpret the pattern of stars flowing across the two telescope fields, comparing it with the known star positions. It would then know which stars were passing across the telescope's sight lines.

So informed, tiny pulses of gas would then be fired, every thirty seconds or so as needed, to tweak the telescope viewing directions so as to point, very precisely, to the stars next in line to be observed. And so on throughout the life of the satellite in space. If and when the satellite was hit by tiny micrometeorites flying through space, the accurate pointing had to recover from these unexpected nudges also.

If this sounds somewhat of a circular logic—star images used to determine the viewing directions, themselves then used to adjust the attitude— it was a concern that we all shared and which we were most anxious, after launch, to see demonstrated. The tiny gas jet thrusters issued tiny puffs of gas, on computer demand, to nudge the satellite just a tiny fraction. They would have to work, reliably, for the full duration of the mission.

If knowledge of the spin rate or the pointing direction were lost, the only way to recover would be to reposition the satellite to its Sun-pointing mode using the coarse sun sensors, then repeat the acquisition steps. The entire procedure would take several hours. It would demand a full turn-out of the operations team—orbital engineers, flight dynamics experts, and instrument operation and satellite attitude groups—and was a delicate and stressful procedure. When first executed, it took over the main control room of the operations centre in Germany due to its size, its complexity, and its criticality.

In the detailed unfolding of the technical design of Hipparcos, the complex attitude control system always remained central to the project's feasibility. That it did work was testament to the expertise of the Matra Marconi Space team at Velizy in France who designed it, ESA project team members Hamid Hassan and Kai Clausen who supervised it, and science team members Erik Høg from Copenhagen and Rudolf Le Poole from Leiden who sat in on various confidential closed meetings with the industrial teams to optimise it for the specific demands of space astrometry.

Le Poole played a key role on the science team. Ten years my senior, he was a bundle of energy. Rarely did he put pen to paper, or in later years, fingers to keyboard. He worked by catalysis. His knowledge of physics was impressively broad, and he had a lightning insight into how things worked and why. He did not design an instrument like Høg could, nor work through the precise mathematics like Lindegren, but he reveled in probing everything, and he made sure that the rest of us questioned any possible missing links. He advised the industrial teams on the attitude control, the grid, the detectors, and the gyroscopes, advised the project team on the alignment and

Rudolf Le Poole (2006)

calibration, sat on the science team, and on the steering committee of the input catalogue and one of the data analysis teams. He laughed a lot, and we talked late into the night when I had problems to discuss. We skated long distances together when the canals froze in Holland, which they did a plenty in the 1980s but rarely thereafter, although a splendid freeze in January 2009 saw us back on the ice together for a reunion. In the wonderful Boerhaave Museum of science in the city of Leiden, sits Philips' first electron microscope, and I spotted that a label on it credits the design to another Le Poole. "Yes," he said, "he was my father." It is interesting how traits propagate down through the generations.

ONE FINAL PIECE OF THE EXPERIMENT which I have side-stepped so far merits examination in a little more detail. We have seen the intricate telescope with its two viewing directions which creates images of the stars at the common focus. And we have looked at the marvels of an attitude control technique capable of spinning the satellite in a carefully controlled manner to point to the stars. The remaining decisive step is how to actually measure the angles between the stars along the circles being scanned on the sky. How to sense and collect the very angles necessary to feed into the processing system on ground which would spit out the positions, motions, and parallaxes of each star measured at the finale of the three year observing period.

Our star surveyor pre-dated CCD technology. The only light-sensitive detectors that could be used were types of photomultiplier tube, which convert light falling onto them into an electrical current. The basic phenomenon—the photoelectric effect—was first described in the opening years of the twentieth century. Explaining what was happening had proved problematic at that time, because the detailed behaviour disagreed with James Clerk Maxwell's wave theory of light. Albert Einstein solved the paradox by thinking of light as discrete particles, or photons, rather than continuous waves. His explanation led to the quantum revolution in physics, and it was this piece of work, rather than relativity, which duly earned him the Nobel prize in 1921.

So it was that photomultiplier tubes located behind the focus recorded the pattern of light bursts as star images entered the telescope's sight line as it scanned the heavens. The images passed across a special asymmetrical slitted mask in the focal plane. This gave rise to an electrical signal which encoded the satellite's directions in space, information interpreted by the satellite's computer, and used to adjust its path along the sky.

In the central part of the focal area was something even more remarkable. Etched onto a three centimeter square mask, Swiss engineers fabricated a tiny grid of nearly three thousand alternately opaque and transparent parallel slits. The grid, so microscopic in pitch that the pattern printed on it was quite invisible to the naked eye, was constructed and laid down with masterful and painstaking precision. The size of a matchbox, it was the result of a difficult and dedicated technology development programme that had extended over more than three years. It was the very heart of the instrument—star images passing across it gave the pulsating signals that encoded their positions.

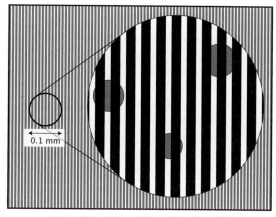

Part of the tiny grid, much enlarged, with star images crossing behind it

As light from the stars crossed this tiny grid, the detector sensed the regular train of peaks and troughs transmitted. Just like looking through a picket fence at a moving light beyond, the source would appear to fluctuate in intensity as it moved, but with a regular uniformity. The positions of the peaks and troughs of different stars gave the information used to pin down their angular separations. Stars from the same telescope viewing field had exactly these separations. Stars from the two different fields differed additionally by the known angle between the two lines of sight. This angle was rigidly fixed by the beam-combining mirror, its own stability ensured by tiny heaters surrounding the telescope providing meticulous thermal control.

Converted to electrical signals, digitised, labeled by the on-board computer, and sent to ground, the data processing teams preparing themselves far below would have all the information needed to re-assemble the pieces of this massive celestial jigsaw.

AS ALL OF THE VARIOUS PIECES of hardware, designed, constructed and tested in various high-tech industries throughout Europe came together, they were shipped to the integration centres in Matra Marconi Space in Toulouse in France and in Alenia Spazio in Torino in Italy. Industrial leaders Michel Bouffard in Toulouse and Bruno Strim in Torino took charge of these final steps. Another jigsaw of hardware pieces was being assembled, final testament to the specifications and detailed design steps which had been set up by ESA and the industrial prime contractors themselves over the preceding years.

And then in they flowed, the synchronised culmination of five years of technological development across Europe: the beam-combining mirror from REOSC Optique at Saint Pierre du Perray; the spherical, folding and relay mirrors from Carl Zeiss GmbH in Oberkochen; the straylight-suppressing baffles from CASA in Madrid; the modulating grid from CSEM in Neuchâtel in Switzerland; the mechanism control system and the thermal control electronics from Dornier Satellite Systems in Friedrichshafen; optical filters, the experiment structures and the attitude and orbit control system from Matra Marconi Space in Velizy in France; instrument switching mechanisms from Oerlikon–Contraves in Zurich; the image dissector tube and photomultiplier detectors assembled by the Dutch Space Research Organisation, SRON in The Netherlands; the delicate refocusing assembly mechanism designed by TNO–TPD in Delft; the electrical power subsystem from British Aerospace in Bristol; the structure and attitude control system from Daimler–Benz Aerospace in Bremen; the solar arrays and thermal control system from Fokker Space System in Leiden; the data handling and telecommunications system from Saab–Ericsson Space in Göteborg in Sweden; and the apogee boost motor from SEP in France. Groups from the Institut d'Astrophysique in Liège in Belgium and the Laboratoire d'Astronomie Spatiale in Marseille masterminded the optical performance, calibration and alignment tests; Captec in Dublin devised the calibration sequences to be run through in orbit, and Logica in London programmed the on-board computer software.

Everything was now on a critical path to launch. There were many tense moments, many crises, many difficult decisions, and many long meetings and too many late nights.

Kai Clausen tells how, in the final stages of integration, he had been summoned to Madrid to hand-carry a replacement part for an antenna which had broken during tests, and which was now needed urgently in Toulouse. He was dropped at Barajas airport with the flight unit reposing securely in a multiple layer metal suitcase, protected from shocks and from any possible contamination. An escort through the airport had even been arranged with the Guardia Civil. The duo marched swiftly and importantly through to customs, at which point Clausen with

Final satellite integration (1987)

no knowledge of Spanish, and the policeman with no knowledge of English, both found themselves at a loss to explain to the vigilant officer as to exactly what each was doing in the other's company. The officer insisted on seeing what was in a case which demanded an armed escort. Clausen, still unable to communicate what was going on, had no option but to resist the officer's strenuous attempts to pick up the antenna part for closer examination. He had, he said later, visions of invalidating the strict cleanliness requirements, a simple finger print which might domino down to a launch cancellation and a penalty fee of many millions. Before the situation got completely out of hand, Clausen produced a fistful of documents in Spanish, which happily included adequate customs declarations thoughtfully prepared in advance.

With these intense and delicate activities carried out in various companies and many countries over several years, the Hipparcos satellite was finally assembled and tested. It was passed from IABG in Munich for tests of the thermal properties, to Liège for further extensive checks and measurements under vacuum conditions, to Intespace in Toulouse for vibration stresses to simulate the vigorous shaking that would be experienced at launch, and on to the vacuum chamber of ESA's Large Solar Simulator at its technical centre in Noordwijk. There it was gently spun at five revolutions per minute under a flood of intense optical and ultraviolet light, simulating the harsh and varying conditions of solar illumination that it would be subjected to in space.

Testing in the Large Solar Simulator (February 1988)

It was during a night time test in this Large Solar Simulator that engineers were alarmed to find the satellite starting to vibrate when the spin rate was lowered. Test facilities were at a premium, and a fix was urgent. No time was lost in calling out mechanical experts from TNO–TPD, fifty kilometers away. They arrived at 3 am, and quickly traced the problem to damaged teeth of the coupling gearbox. In an unparalleled stroke of efficiency, the report was on Hassan's desk first thing the following morning. So too was the bill.

Eventually signed off by industry as all present and correct, confirmed by the ESA project team leader Hamid Hassan, approved by myself as project scientist and through the fiat of the science team, acknowledged as ready for action by the flight operations team in Germany and for launch by the Arianespace authorities, and in turn by an independent launch readiness review panel under the authority of the Director of Science, Roger–Maurice Bonnet, the work on ground was at last completed.

ALL OF US WERE CONSCIOUS that each piece would have to do its work in orbit for the satellite to take its place in history. The most perfectly polished mirrors in the annals of astronomy would have to work together with exquisitely sensitive detectors operating in the harsh environment of space. Featherweight thrusters would gently spin the satellite to survey the heavens. Computers and transponders and solar panels and temperature controls would all need to operate flawlessly. Optics, baffles and filters would have to remain in perfect alignment. There could be no rehearsal once the satellite was put into space, no possibility of adjustments or fixes if anything went wrong.

The satellite was crated, sealed, and shipped to French Guiana. And we pick up the story where we left it in the prologue, after the satellite's successful launch into space on Tuesday 8 August 1989.

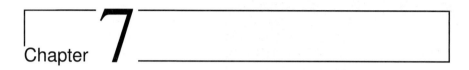

Chapter 7

Disaster Unfolds

*It is impossible to make real progress in technology without gam-
bling. And the trouble with gambling is that you do not always
win.*

Freeman Dyson, in 'Disturbing the Universe' (1979)

T HE IDEA OF A GEOSTATIONARY satellite first appeared in the scientific literature in the early twentieth century, but it was more widely popularised through the insights of science fiction giant Arthur C. Clarke in 1945.

Without a constant source of propulsion to maintain its altitude, a satellite orbiting the Earth must move at a velocity such that its outward centrifugal force matches the downward force of gravity. In low-Earth orbit, a satellite stationed just a few hundred kilometers above the Earth, like the Hubble Space Telescope, will circle around it in a brisk ninety minutes. For the
two dozen GPS satellites constituting the global navigation system constellation at around twenty thousand kilometers altitude, the orbital period slows
down to around twelve hours.

Far above even the GPS satellites, at slightly below 36 000 kilometers altitude, a satellite's orbital period of twenty four hours just matches that of the
rotating Earth far below. A satellite in a geostationary orbit therefore hangs
above the Earth's equator at an enormous height, such that its orbital speed
precisely matches that of the Earth's rotation. As a result, such a satellite is
then poised, apparently motionless, above a point on the Earth's surface.

M. Perryman, *The Making of History's Greatest Star Map*, Astronomers' Universe   125
DOI 10.1007/978-3-642-11602-5_8, © Springer-Verlag Berlin Heidelberg 2010

The geostationary orbit is chosen by commercial communications and broadcast satellite operators because the satellite simply sits strategically above their designated target customers. As a result, large numbers of them now hover high above Earth, all at the same altitude, all following the imaginary line of the Earth's equator far below, and all endlessly pursuing each other in disciplined formation, spread out along their designated orbital longitude. Hipparcos had been granted its position in this procession by the International Telecommunications Union; at twelve degrees west of the Greenwich meridian it would spend its operational lifetime high over the Atlantic Ocean, positioned midway between Liberia and Ascension Island. Here it was to remain for three years, and from its spectacular vantage point in space it would be in direct and continuous contact with the fifteen-meter radio antenna of the European Space Agency's ground station network in the Odenwald, near Darmstadt in Germany.

A geostationary orbit had been selected for the Hipparcos satellite for a number of compelling reasons. First, in this location it would be in direct radio communication from the single ground station for the full twenty four hours each day. Commands could then be sent continually up from the ground and, more importantly, the relentless stream of science data could be sent down by radio link to a single huge antenna dish pointed in its fixed direction. Second, this location would provide an uninterrupted view of the depths of space, and give the satellite a grandstand view of the stars that it would be surveying. Third, from its location high above the atmosphere, the gravity and thermal forces wrestling with the exquisitely sensitive instrument would be miniscule, consigned to irrelevance.

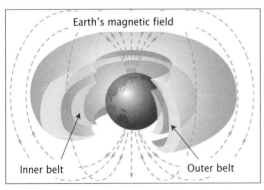

The Van Allen radiation belts around Earth

The geostationary orbit also lies above the Earth's Van Allen belts, vast toroidal sweeps of energetic charged sub-atomic particles, held in place by the Earth's magnetic field. Related to the magnificent polar aurora, where particles ejected from the Sun are funneled by the Earth's magnetic field, strike the upper atmosphere and shine by fluorescence, energetic protons trapped in this inner belt would slam unimpeded into any satellite unfortunate enough to be in its vicinity. The radiation would be highly damaging to its electrical solar cells, and potentially fatal to its sensitive electronic circuits.

A launch vehicle is not able to place its satellite cargo directly into a circular geostationary orbit at high altitude, but only into a very elliptical 'transfer' orbit. In the same way, a croupier can't launch a ball into its circular track in a roulette wheel merely by flicking it outwards from the centre. But what the transfer orbit does is to connect the final push of the rocket with the geostationary orbit far above.

Jettisoned by the rocket into this transfer orbit, the satellite rushes down towards Earth passing within only a few hundred kilometers of its surface, before swinging round it and hurtling back out to its farthest point high above, its apogee, at 36 000 kilometers altitude. This extravagant sweeping orbit is repeated every few hours. It's a stable enough orbit, apart from aerodynamic drag at its closest approaches to Earth, but neither particularly useful, nor one convenient for typical operations.

Satellites destined for geostationary orbit are circularised into the final orbit by a powerful blast from a special motor built onto the satellite itself. Ignited at its apogee point furthest from Earth, through a carefully-timed command from the ground station, this transforms the ten-hour elliptical transfer orbit into its intended, circular, twenty four-hour geostationary track. Thereafter, Hipparcos would be gently nudged, using its own thrusters, to its parking position twelve degrees west of the Greenwich meridian, there to spend the rest of its life, scanning the heavens.

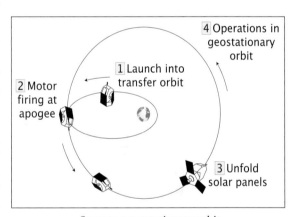

Steps to a geostationary orbit

THIRTY SEVEN HOURS after a perfect launch, and precisely as designed, planned, and minutely rehearsed, the satellite operations team in Germany, under the authority of its leader Howard Nye, and supported by project team and industry engineers led by Kai Clausen, issued the command to ignite the satellite's apogee boost motor to circularise its orbit.

There was no response. Motion trackers on the ground could sense no change in the satellite's elliptical trajectory. And there was only one explanation: the boost motor firing had failed. Shock and consternation greeted this wholly unexpected development.

ALTHOUGH THE MOTOR FIRING COMMAND would be re-issued from the control centre on ground at subsequent apogee passes, there was already little optimism that future attempts would have a different outcome. Like starting a problem car, if the ignition doesn't fire first time, it's probably not going to start the next—and raising the bonnet to tinker inside wasn't an option. The most urgent and troubling concern was that each unsuccessful attempt would consign the satellite to yet another elliptical orbit swinging around the Earth, each one taking it through the deadly grip of the radiation belts. The satellite's sensitive electrical circuitry had been designed to survive only a few of these Van Allen belt passages before its escape to its geostationary haven. Time was already running out.

The failure of the first boost motor firing was announced by mission control in Germany to the Kourou launch centre shortly before most of those who had shared in the excitement of the launch climbed aboard their flights back to Europe. Spirits were crushed. The jubilation which we had felt nearly two days before evaporated as the unjust turn of events hit home. The boost motor is a routine satellite component, and failure is all but unknown. By the same token, however, each is a single-shot pyrotechnic device which can hardly be tested on ground. The Mage-2 motor, with its half a tonne of solid propellant, should have burned through in under a minute and lifted Hipparcos to its operational orbit high above the Earth. But it remained stubbornly inert, like a jumbo firework whose fuse had burnt through but which had failed to go off. It is common practice in space missions to avoid what are called single-point failures, in which the failure of one component would otherwise lead to total mission loss. This critical part was no exception—a second, backup, firing chain was built in, but this too failed to ignite.

Investigations hoping to diagnose the reasons, and further attempts to activate it, would be made, but prospects appeared inordinately bleak. Stranded in its highly elliptical transfer orbit, it seemed clear that Hipparcos could not escape its daily devastating encounters with the Van Allen radiation belt, and that it would surely be destroyed within a few days or weeks at most. This innovative and expensive experiment seemed destined not to perform its measurements. Whether this would be a definitive end, with a possible replacement neither funded nor supported in view of new priorities in Europe's advancing and highly competitive space science programme, or whether, in the most optimistic scenario, it might represent a hiatus of several years during which a replacement could be financed and manufactured, seemed irrelevant. A second, spare satellite is rarely constructed, nor is launch insurance taken out, in view of the cost-to-risk analysis. Both prospects seemed acutely distressing, both equally catastrophic. The high-stake gamble of careers and science advances had not, it seemed, paid off.

It was with bitter resignation and funereal silence that the teams, so elated two days before, ascended the aircraft steps, and settled into their seats and their private thoughts for the despondent flight home. I discussed the dismal prospects with project manager Hamid Hassan in the baking heat of the refueling stopover at Dakar in Senegal. Arriving in Paris after midnight, and with time at an essence, we took a charter flight on to Frankfurt, and joined the satellite operations team in Darmstadt as day broke across Europe.

The reception committee included the head of the spacecraft section in Hassan's team, Kai Clausen. Ten hours after leaving Kourou, we were sitting in a small conference office next to the imposing main control room, receiving a complete debrief of recent events and the current, dismal, prognosis. Normally of jovial demeanour, masking an extensive space technology knowledge, Clausen was haggard and deeply troubled. In the cold light of day, the prospects seemed bleaker than ever. We settled down to debate the options ahead.

Andrew Murray, a leading member of the science team who had watched the launch in Kourou, wrote to me two weeks later: "We must hope that whatever can be salvaged from the present mission will provide ample justification for a future one. We have had cruel ill luck, but if we can secure support for another mission within a few years, the delay will be seen as small in the context of the scientific aims for which astronomers have been striving for three hundred years." I appreciated his words, and admired his resolve, but found it difficult to share his optimism.

Another letter of commiseration, from my head of department, Dr Edgar Page, ran: "I know only too well how it feels to invest years in a project and see it all taken away," a reference to his involvement in the ESRO II mission which dropped into the Pacific Ocean following a Scout launch failure in 1967. We had each learned first hand that the unexpected collapse of a lifetime's work is keenly felt.

Pierre Lacroute, even at 83, remained stoical. "It doesn't really matter," he uttered on hearing the news in Kourou, recalled by Jean Kovalevsky twenty years later. "It will be relaunched in a few years."

Over the following days and weeks, Hipparcos figured large in the world's scientific and popular media, but not at all for the reasons we had hoped. A well-attended press conference was hosted at the operations centre in Darmstadt on 17 August 1989. Lined up on the podium, like plastic ducks at a fairground shooting range, were me, Hamid Hassan, and Roger–Maurice Bonnet, along with acting director of satellite operations Wilhelm Brado, flight operations director David Wilkins, and the general manager of Matra Marconi Space, Claude Goumy.

ESA

Press conference, Darmstadt, August 1989

I played my part in explaining, probably inadequately, how two hundred scientists across Europe were taking the satellite's imminent demise. We could not help viewing the press, albeit rightfully curious, as well as courteous and well-briefed, as vultures picking over the bones of a carcass already pronounced dead. Jon Darius, an enthusiastic and knowledgeable correspondent for *New Scientist* magazine who had been at the launch, summarised the situation ten days later[36]:

> The fate of Hipparcos, Europe's latest scientific satellite, hangs in the balance. After what looked like an excellent start, the satellite has run into severe problems. The apogee booster motor, the rocket that was to have taken the satellite into its correct orbit, failed. So far, there have been three unsuccessful attempts to fire the motor: a fourth attempt was due on 17 August. Unless engineers can solve this problem, the satellite will be unable to attain the goals of the mission, either in terms of the number of stars it surveys, or the accuracy of its measurements.

© Independent 1989

Scientists losing hope for satellite trapped in orbit

SCIENTISTS are losing hope that they can save the Hipparcos satellite, which was to be one of the most important science projects of the century, from being trapped forever in the wrong

By Mary Fagan
Technology Correspondent

ground station at Perth in Australia, but even then the satellite

The Independent, 15 August 1989

(Mary Fagan)

Project manager Hamid Hassan proposed a priority of putting the satellite into its slowly spinning state, with the sole objective of demonstrating to everyone's satisfaction that the innovative and intricate scanning of the sky could work as planned. Confidence in this tricky manouevre would at least remove one operational concern when lobbying for a replacement mission. The operations team brainstormed to figure out if anything else could be salvaged. Engineers at ESA's research and technology centre in The Netherlands rallied around. They began the lengthy task of analysing the on-board voltages and currents to evaluate the radiation effects already beginning to cripple the life-sustaining solar panels.

The science advisory team, which had guided the mission from its conception ten years earlier, convened for an emergency meeting in ESOC on 16 August and demanded in writing to the Director of Science that the agency immediately put into gear the funds and plans for a replacement. But finding the necessary hundred million dollars in an already over-extended programme would be an uphill battle. The money could only come from other projects which would themselves then suffer delays. Nobody believed that prospects for a replacement were good. The mood remained despondent.

AN INDEPENDENT FAILURE ENQUIRY BOARD was set up to investigate the catastrophe, chaired by French defence department missile expert Daniel Reydellet. Later to hold the rank of four-star general in the French armament corps, Reydellet was a relentless investigator who had shouldered responsibility for the solid rocket motors for the French nuclear deterrent forces, and would later assume charge of their entire armament programme. After ten months of meticulous probing, including visits to relevant industries and interviews with key managers, the board reported its findings at a press conference at ESA's Paris headquarters on 6 July 1990[37].

They pointed the finger firmly at a part of the firing circuitry called the through-bulkhead initiator, a matchbox-sized component that delivers hot gas to the igniter of the motor's solid propellant. Manufactured by California company Hi-Shear, it was one of the very few components of the satellite manufactured outside Europe. The investigators were simultaneously critical of the French company SEP, who had assembled the entire ignition system, citing "insufficient diligence" in checking their sub-contractor's work. For their part, Hi-Shear officials said that they were not aware that the initiators sold indirectly to SEP in mid-1988 were to be used in space. Claude Goumy, president of Matra Marconi Space, countered that the component "was a part of the

Daniel Reydellet (2007)

[space] shuttle and designed specifically for space applications", while SEP said that they had relied on NASA's seal of approval for the Hi-Shear initiators given in the mid-1970s.

The findings were particularly bitter for a most unfortunate reason. The pyrotechnic devices actually came with a 'use-by' date, to ensure that the explosive charge would not degrade with time into an unreliable state. In a master stroke of ill-fate, both original devices had been replaced with newer items while the satellite had sat in storage for several months, a year before launch, while unrelated problems with the Ariane rocket family were being investigated and resolved.

From the mountain of project documentation, the enquiry board traced the serial numbers of the two Hipparcos firing circuits—99822 and 99823. Engineering teams tracked down the next in line, still unused, and proceeded to test fire it. It failed. So too did another from the same batch, 99830, during tense subsequent tests on 6 June 1990. The board concluded that the manufacturing sequence used in that particular batch was faulty. The original igniters—which probably would have worked—had actually been replaced by substandard components.

Hi-Shear Part Blamed For Hipparcos Failure

By PETER B. de SELDING
Space News Staff Writer

PARIS — A faulty component supplied
by a California company was the principal

There remains a possibility that the explosive transfer lines connected to the initiators might themselves have delivered a
weak charge which, when combined with
initiators of marginal quality could have

Space News, 16 July 1990

(Peter B. de Selding)

The cause of the disaster established beyond reasonable doubt, Jean-Luc Lagardère, CEO of the massive Lagardère media and defense conglomerate which included Matra Marconi Space, invited Roger–Maurice Bonnet to his headquarters near the Place de l'Étoile in Paris. In a most civilised gesture, he solemnly presented apologies on behalf of his company to both ESA and to the scientific community.

The fastidious quality assurance procedures, a crucial part of disciplined space ventures, were subsequently re-written to prevent such recurrences. Stringent conditions were imposed on items procured from a single manufacturing batch. And the enquiry board recommended that European authorities should no longer take a NASA seal of approval as unqualified proof of reliability. This was all part of the inexorable consolidation of space experience. The lessons, nonetheless, offered but nugatory solace to the Hipparcos scientists.

EIGHT YEARS LATER, the monumental Hipparcos star catalogue would, in fact, be published. It would exceed all expectations, surpassing the specifications established nearly twenty years previously. It would provide the most wonderful stereoscopic view of the stars in our neighbourhood of space, and an unprecedented and plentiful scientific harvest for astronomers for decades to come. It would alter our understanding of our Galaxy, and our place within it.

All of this was unknown at the time. The boost motor never fired, but what unfolded over the ensuing months and years instead represented a triumph both of technical and scientific ingenuity and of relentless persistence.

Chapter 8

Mission Recovery

*The greater the loyalty of a group toward the group, the greater
is the motivation among the members to achieve the goals of the
group, and the greater the probability that the group will achieve
its goals.*

Rensis Likert, in 'The Human Organization' (1967)

WITH THE LAUNCH ON 8 AUGUST 1989, the pivotal contribution of
Arianespace was completed. At the time of the unsuccessful first
attempt to ignite the apogee boost motor two days later, the Hipparcos
scientists and engineers were distributed in a number of places.

Most of the ESA project team, under project manager Hamid Hassan,
had been in Kourou, supervising the activities of the Ariane team in their
preparations for launch. With the satellite components delivered and integrated, the prime contractor teams of Matra Marconi Space and Alenia were
now back in their home countries, and mostly already dispersed to other
projects. The leaders of the various scientific teams responsible for the science aspects had also attended the launch in Kourou. They duly returned to
Europe, and back to their scientific institutes. If things had gone according
to plan, they would now be ready and waiting for the first of the science data
to come streaming down from the satellite.

The principal actors in the next phase of the programme were the operations group stationed at ESA's operations centre, ESOC, in Darmstadt in
Germany. A team of around twenty had been gearing up to their substantial
role in the project for the last three or four years. Under the responsibility of

M. Perryman, *The Making of History's Greatest Star Map*, Astronomers' Universe
DOI 10.1007/978-3-642-11602-5_9, © Springer-Verlag Berlin Heidelberg 2010

spacecraft operations manager Howard Nye until just after launch, responsibility was thereafter transferred to Dietmar Heger. The team swelled to more than a hundred for the activities around launch, stabilising to around sixty for the revised operations over the next three years.

ESA's main control room at ESOC

The team had been responsible for setting up all operational procedures, in close consultation with the industrial teams, the ESA project team, and Arianespace. Here were assembled the experts in satellite operations, orbital manouevres, data transmission, and instrument monitoring. Here, all of the complex operating procedures of the Hipparcos satellite had been planned, documented and rehearsed over the previous months. Procedures for everything we could think of were in place: shift planning, instrument calibration, data verification, satellite distance ranging, command uplink, data downlink, checks of the star observing list, procedures for satellite spin-up, spin-down, coarse pointing, fine pointing, and plenty of others. Contingency procedures were documented and filed. All the actors knew their place on the stage, and everyone knew their lines. None of the team were at the launch in Kourou—their place was in Darmstadt, ready to take control of the satellite as soon as it was ejected from the launcher's nose cone, hurtling on its way across the Atlantic Ocean towards Europe. Yet the launch was the highlight in their diaries, and the operations centre, like the other major establishments of ESA—Noordwijk in The Netherlands, Paris in France, and Frascati in Italy—hosted ESA staff and members of the press so that they, too, could watch as it happened.

THE FIRST WEEKS OF OPERATIONS were to be conducted in the impressive main control room of the operations centre, with its own giant screens, status lights, and banks of control consoles, all manned for launch by the various flight engineers and instrument controllers.

Stepping into the main operations control room never failed to impress. A large hall, some twenty meters across, it was always lit serenely with subdued lighting, always hushed with damped acoustics to lend an aura of isolation, a world cocooned from the one outside. On the front wall, screens with continuously refreshed maps showed satellites being followed and daylight tracks on Earth, panels showing times and dates and status counters, always ticking, always something happening. Manuals and emergency procedures lined the back walls. Serried banks of computer consoles were set out for the satellite controllers, as well as the telemetry engineers, attitude experts, thermal engineers, ground station controllers, and mission analysis experts. Commands came in over loudspeakers and out through microphones. All this to keep an object the size of a car alive high above.

If the nerve centre of the operations was in Darmstadt, the eyes and ears of the team were more widely scattered around the world. The main antenna receiving word from the satellite was a fifteen meter dish fifty kilometers outside Darmstadt, in the rolling hills of the Odenwald. Other smaller antennae were distributed around the world. Particularly critical just after launch, when the satellite passes close to Earth, were a network of relay stations

Odenwald ground station

around the equator, all linked to the main control centre in Darmstadt— radio receivers at the Kourou station itself, as well as Malindi in Kenya, and Perth in Australia. Data from the satellite would come down to these various listening stations, to be piped back to Darmstadt through dedicated telephone lines. There, in customised computer banks, the pieces from the different antennae would be assembled, time stamped, checked for integrity, and inspected for quality. Copies would be made for immediate despatch to the processing teams around Europe, another backed up in a secure vault far away. Each and every second, twenty four thousand bits of data encoding the mysteries of the Universe would be arriving from space.

Commands from the control centre to the satellite would follow a similar path in reverse. Up would go the next piece of the star catalogue to set the scanning for the next few hours, tweaks to the telescope focus, and adjustments to the thermal balance. All commands would be checked and double checked before leaving Earth. Another sharp tool in the ground station box of tricks were the ranging commands. At regular intervals, radio signals launched up and bounced straight back down from the satellite gave its distance. From a collection of values over a day or so, the mission analysis experts reconstructed the satellite's orbit. This information, too, was needed by the scientific teams constructing their great star map.

THE OPERATIONS TEAM had issued the command to fire the apogee motor from the main control room in Darmstadt, and this was to remain the centre of operations for the next three and a half years. The days after launch would be spent analysing the satellite data stream, checking the circuitry, and trying to come up with an explanation of what had gone wrong.

Kai Clausen (2008)

As in forensic pathology, the quality of evidence that could be extracted from the crime scene was rather surprising. Even more so in this case given that the body was tens of thousands of kilometers away. Kai Clausen supervised a reprogramming of the onboard computer to send down the current readings in the detonation circuitry throughout subsequent boost motor firings commands, tagged with a very fine time resolution of one thousandth of a second. They concluded, in a report presented to the failure enquiry board, that the relay circuitry was functioning as planned, and that it was the through bulkhead initiator that must be at fault.

Board member Dermot O'Sullivan was tasked, explicitly, to investigate whether the satellite had been sabotaged, whether someone might have cut through the boost motor firing line during the satellite's integration or its preparation for launch. A careful examination and analysis of the detailed circuit diagrams proved to everyone's satisfaction that this could not be the explanation—if the line had been cut, the solar array panels would not have opened since their own pyrotechnic firing commands ran through the same wiring harness.

Other attempts were made to fire the motor after that first fateful command: on the third day after launch; once on the fourth with the forensic tests; once on the ninth after slowing the spin rate; and finally seventeen days after launch after heating the motor by facing it towards the Sun for several days, haunches to the hearth.

In an exercise considerably more complex than it might sound, the precise moment of firing had first to be calculated, according to the time that, in its ten hour orbit, the satellite would just be turning back from its furthest point from Earth. This could be at any time of the day or night. The countdown sequence was complicated because of gaps in the ground coverage, and certain other commands that had to be issued first. Each enactment of the drama accordingly started around twelve hours before the actual firing time. After the first few attempts, the operations team had been pretty much awake around the clock for several days. After my arrival from Kourou I joined the team in the hotel lobby one morning for a 4 am start. Coffee had been put out, but that was all that was on offer by way of breakfast. The dozen members of the project launch team, led by Clausen, were bleary-eyed, groggy and depressed, the faces of a thoroughly dispirited army that had been beaten in battle. It did not feel like much to show for ten years of work. We set off through the remains of the night for another attempt.

The first week or so after the apogee motor firing failure were particularly tense and especially uncertain. Project manager Hassan's formal responsibility would have come to a close at the end of satellite commissioning, which had been planned for some thirty days into the mission. At that point the instrument should have been fine tuned, ready to start its survey of the sky. Instead, the boost motor failure appeared to signify the imminent demise of the satellite.

Interest in the satellite was flagging. I offered to take over the troubled mantle of project manager. Most had given up hope for anything much to come from the crippled satellite anyway. Although it was unprecedented for one person in ESA to hold simultaneously the dual roles of project scientist and project manager, these were

HIPPARCOS

Bad day for astronomers

Kourou, French Guiana
AFTER a perfect launch in French Guiana on 8 August, the European astronomy satellite Hipparcos ran into trouble when the apogee motor failed three times to respond to commands to fire. The motor is responsible for lifting the satellite into geostationary orbit around the Earth. As carrying a West German direct broadcasting television satellite, TV-SAT 2, as well as Hipparcos. The first launch attempt was stopped seven seconds short of ignition by an overzealous computer. But then the mission got off to a good start with the perfect positioning of Hipparcos and TV-SAT 2 in their transfer orbits. Just

Reprinted by permission of Macmillan Publishers

Nature Magazine, 17 August 1989

unusual times demanding appropriate measures, and nobody doubted that the one person driven to get the most from the ailing satellite was the chief scientist in charge. The Director of Science, Roger–Maurice Bonnet, equally anxious to get the most from it rather than have to dig deep into the budget for a replacement satellite, signed off my authority. Bonnet remained ever hopeful that Hipparcos would somehow perform, but whether through insight, optimism or desperation I was never really certain.

ON 26 AUGUST 1989, eighteen days after launch, we finally and formally abandoned hope of a successful motor firing. I sat with Kai Clausen, ground segment manager Jozef van der Ha, and spacecraft operations managers Howard Nye and Dietmar Heger to examine our options, agreeing that all effort would be henceforth dedicated to designing and optimising a formal recovery plan. Clausen had been the head of the spacecraft section of the project in ESA, and remained fully committed to do everything possible to get something from it. The objective was to wrestle whatever we could from the satellite, by whatever means, before it died.

First priority was to make use of the explosive hydrazine propellant fuel on board, which was to have been used for the final station-keeping manouevres, to raise the satellite's point of closest approach to Earth as high as possible above it, in practice a little more than a mere five hundred kilometers. In turn, this would help in reducing the extra forces which would be hitting the satellite as it slammed through the upper atmosphere on its closest approach to the Earth. It would also lengthen the orbital period by a small amount, to just over ten hours. Orbit expert Alain Schutz masterminded the timings and durations of the hydrazine firings, and the orientation of the satellite in space, so that it would move up in the correct direction, and not off in some other.

Next was to go from the stable rapidly spinning state left by the launcher to the slowly spinning demanded by the star observations, a transition which was complex and irreversible, made considerably more tricky by the absence of permanent ground coverage. The most critical step of them all was the unfolding of the three solar panels. Due to the limited ground coverage, all procedures had to be redesigned, and new contingency plans drawn up.

To understand the rapid damage being inflicted on the solar arrays due to the satellite's repeated passage through the Van Allen belts, ESA's solar array experts quickly got involved in a detailed modeling of the voltage and current diagnostics from the circuitry on board. Over the subsequent weeks and months, Roy Crabb, Geoff Dudley, and Horst Fiebrich were to play the role of hospital consultants in an intensive care ward: scrutinising the numbers, plotting their trends, organising urgent laboratory testing, and building up physical models. Their continuously-updated analysis was nervously awaited by the operations team. I forwarded a weekly summary and prognosis to the science team, to my management, and to the ESA's advisory groups, all of whom were anxious to know how long the patient could be expected to live.

Scientists battle to revive star satellite

By Steve Connor, Technology Correspondent

SCIENTISTS will try to revive a satellite today which is orbiting aimlessly around the Earth after its booster rocket twice failed to respond

The booster rocket, called the apogee engine, yesterday failed to respond to a command to fire which should have moved the satellite into a much bigger

Daily Telegraph, 12 August 1989

(Steve Connor)

I organised and received weekly reports of the activity of the Sun, compiled by Pierre Lantos of the Paris Observatory using their solar telescopes at Meudon. We pored nervously over these, studying the electron and proton doses delivered by the outpourings of the solar wind into the Earth's upper atmosphere. It was the unpredictable streaming of these fateful sub-atomic particles that replenished the Van Allen belts, and which would slowly destroy and eventually kill off our charge. The Van Allen belts were our particular bête noire, the ugly radiation traps where high energy particles heading for Earth are intercepted by the Earth's magnetic field. Our satellite would plunge through them four times a day.

Going on in parallel was a scramble to recruit other big listening dishes into the ground station monitoring network. If Hipparcos had reached its intended orbit, just the one antenna planned, at the Odenwald station near Darmstadt, was all that was needed. It would simply remain in contact with the geostationary satellite for twenty four hours a day. But consigned to its elliptical orbit of just over ten hours' duration, it was only in contact with Odenwald intermittently during the day. For the rest of the time, it was racing around above other parts of the Earth. We needed more ground stations, and we needed them urgently.

Working against us were the numbers: there were only a few suitable antennae spaced around the world, and they could already be committed around the clock anyway. Working against us was time: agreements to plan and make use of these dishes typically took months. Working against us was money: I had a fixed operations budget to work with, planned by ESA management, and approved by the advisory committees. Financial authority is carefully regulated in such a large and complex organisation, and arguments that we were immersed a crisis held no sway with the teetering hierarchy of regulators; procedures had to be followed. So asking for the extra sums needed to defray costs of the additional ground stations would take time also. Like going to the bank to plead for another loan for a failing business, the outcome would be uncertain.

But people can come together in adversity, and here was to be a fine example. Roger–Maurice Bonnet gave his approval to move on the finances. Formally we would need to think of it as a shortening of the operational phase, and we would have to come back to negotiate extensions later if operations lasted that long. More impressive still was the speed with which the operations team, now under Dietmar Heger, moved to engage and equip the small Galliot antenna in Kourou, and a large antenna in Perth, Australia, both part of the ESA tracking network. By mid-September, Perth was fully operational in its support of Hipparcos, and contact with the satellite thereby raised to just over sixty per cent of the time. A further dish in Villfranca outside Madrid was brought in as a backup to the critical Odenwald station.

An important addition was provided across the Atlantic. In a splendid piece of goodwill swiftly implemented, NASA, and engineers at their Goldstone station in the Mojave desert in California, promptly made one of their 34-meter Deep Station Network dishes available to the ESA network for several hours each day. Naturally, communication protocols differed between the two space agencies. The necessary command and reception equipment was quickly procured, put together in Germany, and shipped to Goldstone in early 1990. The high-speed data interfaces gave more problems than expected,

NASA's Goldstone antenna

and it was mid-May before the Mojave desert station was linked in, up and running. But the extra data each day made all the difference, and their giant dish was linked into our network for the next three years. Formally made available on a *quid pro quo* basis, NASA called in the Goldstone debt some years later. That was fine also—as they say in the space business, there's no such thing as a free launch.

Schedules were drawn up as to which hours on what days we could switch each of the various dishes into the much-needed support network. Satellite communication time was raised to ninety per cent, with loss of contact times down to below an hour and a half. But the revised ground system was always awkward to schedule and operate, significantly more complex than would have been considered acceptable during the design phase.

The next step of the recovery plan was the most taxing, and the most crucial, and was embarked upon with no guarantee of success. At this point, Hipparcos was not able to scan the sky. The attitude control system had been worked out for the geostationary orbit, in which the only perturbing forces influencing its motion were due to solar radiation pressure. This is a small but significant force pushing on the solar panels by sunlight incident on them. Correcting thrusters were sized and tuned to balance out this force alone. In its new unplanned path, the forces were much larger, and varied wildly around the orbit. There were extra tugs from the Earth's gravity field as the satellite plunged towards perigee, additional radiation pressure due to light and infrared radiation reflected from the Earth, forces due to the Earth's magnetic field and, as the satellite hurtled around its lowest perigee passes a few hundred kilometers above its surface, atmospheric pressure drag.

On top of all this was the fact that, even with the new antennae in the monitoring network, there were still large gaps of up to an hour or two in the antennae coverage, mainly around perigee. It was going to be a daunting task to redesign the monitoring and control procedures on the ground, and the autonomous propulsion system on the satellite, to handle this massive change in the spacecraft's environment. Much of the flight operations procedures had to be rewritten. Most precarious was when the satellite was out of contact of any of the antennae as it raced around the lowest approach to Earth. During this out-of-contact time, commands still had to be executed on board—opening or closing of the shutters to avoid light overload due to occultations by the Earth or Moon, updating the models of the forces, and so on. All these time-critical commands had to be prepared in advance, uploaded to the satellite, and placed in a buffer for execution at the correct moment. For the first passes, we just held our breath as the satellite disappeared from contact, spluttering and gasping with relief when it emerged under some semblance of control an hour later.

DIETMAR HEGER FROM THE OPERATIONS SIDE, and myself from the science side, continued our regular conferences to draw up the plans for the struggles ahead. Heger, ever resilient, ever cheerful, and determined to succeed, drew on the skills of a talented and committed operations team, in which Alastair McDonald and Alain Schutz played leading roles in the recovery. Kai Clausen assisted in sketching out the changes that would be needed to the on-board software. I drew up the specifications and new contracts for the necessary support by the software team within Matra Marconi Space that had coded up the satellite software when it was still on ground. Only they knew what could be changed now that it was in space.

Very significant was the expertise of Floor van Leeuwen, a Dutch astronomer based at the Royal Greenwich Observatory. Van Leeuwen was responsible for the pieces of the software processing on ground to do with the spinning motion of the satellite. He had acquired an unparalleled understanding of the attitude control system—the perturbing forces, the vagaries of the on-board gyroscopes, and the complexities of the gas jet thrusters. Together, piece-by-piece over the coming weeks, we slowly built up and practiced the procedures to operate the satellite in its new, highly inconvenient orbit. Continually rolling shifts of satellite operations teams, regular feedback from the solar array experts, flexible support from the ground station antenna teams, advice from industry and ESA experts, all had to be juggled.

Steady advances were punctuated by numerous setbacks and false dawns but, finally, useful astronomical data started coming in. Operations had moved to a small dedicated control room, with operators manning the consoles, issuing commands, and verifying the data around the clock. By the end of 1989, three months after launch, it was still only arriving in relatively short stretches of an hour or two at a time, but as we moved into the new year, it increased by leaps and bounds as the new procedures were mastered.

Just when we thought things couldn't really get any worse, Dietmar Heger and Alain Schutz, our expert on the orbital dynamics and scanning motion, slipped a graph onto my desk. Schutz was a man of few words, who we relied upon implicitly for his analyses, which were always perceptive, always clear, and always correct. "There's a big one in March", was all he said. I looked at the title of the graph which was 'eclipses', the vertical axis which was 'duration in minutes', and the horizontal axis which was 'month'. For about twenty minutes each day, the Earth would occupy a position between the satellite and the Sun. This meant that no sunlight was falling on the solar arrays, no power would

Dietmar Heger (1997)

be flowing from them, and the satellite would have to function for those few minutes on battery power alone. These eclipses were known about, and it was why the satellite had batteries.

But Schutz had been busy calculating the eclipse lengths with the new orbit. Adding the new ten-hour elliptical orbit, to the twenty four hour rotation of the Earth, to the Earth's elliptical orbit around the Sun gave the results: some splendid intervals without any eclipses at all, in May and June of 1990 and, much later, in November 1991. But there were a couple of heavily shaded areas where the eclipses would be much longer than originally planned. For these precarious moments, no life-giving sunlight would be sustaining the satellite's energy demands.

The worst of all would hit in March 1990, less than six months hence, at almost two hours duration. If the satellite survived until then, we would have to power down everything not absolutely critical to the satellite's survival, and ride out the long eclipses with the entire instrument in hibernation. Another big headache, more procedures to write and test, and another touch-and-go few days. With more good fortune on our side, the satellite duly survived way beyond March, the eclipse-hibernations also, even if by a worryingly slender margin.

THAT THE UNCERTAINTY ON the satellite's life expectancy persisted for months, and in reality throughout the following three years, resulted from a combination of factors. First amongst them was the uncertainty in the way in which the solar arrays would ultimately fail: the drop in the voltages and currents were flattening off, but still creeping ever closer to their death zone at 54 volts. Once the voltage delivered by a battery drops below a certain critical number, it ceases to serve any useful purpose, and our space arrays were no exception. Whether they would continue to deliver power for the next few months remained an impossible judgment to call. Part of this uncertainty was itself linked to the indeterminacy of the Sun's activity, to the unforeseeable nature of the large solar flares which congregate around solar maximum on their eleven year cycle. Unpredictable eruptions spew out large doses of lethal high-energy protons, heavily augmenting the steady degradation imposed by the trapped Van Allen radiation belts. It was Russian roulette.

The difficult operations continued for the best part of three years, hit by many setbacks, and as many hard fought recoveries. Round the clock shifts continued without remission. After six months we knew we had enough data from the satellite to have a map of the sky, positions only, good to about ten milliseconds of arc—still far off from the eventual goals, but much better than ever done from Earth. After a year, we had our first hints of parallaxes from the ground processing. Eighteen months in, and the star motions started to appear in the data too. Two years of data, even after a few large solar flares and the voltages falling close to their limit, we could see the finishing line. The data were of superb quality when we were able to acquire them in long stretches.

We bit our nails when the solar flares struck, when a ground station dropped temporarily from the network, or when the satellite lost its attitude control and reverted to its emergency sun pointing. We had countless meetings, innumerable updates and adjustments to the on-board software, and much head-scratching to figure how to keep the satellite alive as it slowly began to fall apart. Finally, the radiation damage slowly began to exert its deathly toll on other satellite parts besides the solar arrays. First to fall victim were the crucial gyroscopes which sensed and controlled the ever-evolving forces pushing on the satellite. Later, the thermal control sensors and controllers began to fail. The temperatures within the telescope began to drift beyond acceptable bounds. In each case, redundant systems were switched in as the primary systems failed.

As the last of the redundant gyroscopes came into the control loop, the operations team designed a clever procedure that would run using just two gyroscopes to sense its orientation in space, rather than the three which had been planned. It was a fine battle won by the engineers which kept the satellite spinning and the data flowing for crucial months, but the radiation belts inevitably won the war. In the final days, the communications system itself started to fail.

Cheered on by the science team and their two hundred colleagues in the scientific consortia, by the wider astronomy community which had had its eyes on the results for years, and by the Director of Science, Roger–Maurice Bonnet, the satellite had lived on, even beyond its original design lifetime. After a little more than three years of successful data gathering, by early 1993 the systems were failing catastrophically one by one. As if it had fought valiantly to maintain the data flow so desperately needed to assemble the final star catalogues, the on-board computer finally expired, refusing to restart, ravaged by its prolonged exposure to hostile radiation.

Science team and operations team, ESOC (Darmstadt), 13 July 1993

All further attempts to communicate with the dying satellite failed. Hipparcos, finally, signed off.

Chapter 8

THE SATELLITE HAD BEEN a terribly awkward customer for Dietmar Heger's operations team. But it had also provided its own intellectual challenges for their talented minds. Alastair McDonald, a member of the flight dynamics team, wrote to me after switch off: "Although nobody would have wished it on the mission, the continual problems, from the apogee motor to the on-board computer, provided a constant source of excitement and motivation for everyone. Apart from the phone calls at three in the morning which I do not miss, it was also tremendous fun."

Quite curiously, but most fortuitously, despite the wrong orbit and all the problems that followed from it, the final Hipparcos catalogue would eventually exceed all expectations, surpassing the specifications established nearly twenty years previously. Just as in an aircraft, or in a public building, technical safety margins built into the design eventually worked together in our favour, and more than compensated for the horrendous complications of the unintended orbit. And over the next few years, the scientific teams would work zealously to wring all possible improvements in accuracy from the complex data stream.

Meanwhile, the satellite operations work was done. The experts moved on to new projects, more experience under their belts. The Hipparcos control room was taken over by the next project, our own operations manuals and emergency procedures and log books thrown out to make way for the next. Life moves on. Hipparcos was history. It was a sad moment to accept that this brave fighter was no longer with us. But none of the team would ever forget their three years of intensive care for their valiant patient.

Chapter 9

Science in the Making

Creative minds never get much of a mention. It's warlords, politi-
cians and other confidence tricksters who get talked about all the
time, not people with creative minds.

Willem Frederik Hermans, in 'Beyond Sleep' (1966)

I N THE YEARS LEADING UP TO THE LAUNCH, three separate scientific teams
had been preparing for the flood of data which would be beamed to Earth
once the satellite was in orbit. An almost interminable three-year long binary string of ones and zeros would be transmitted from space as the satellite turned slowly across the sky. The data stream would be cleverly labeled
to signify all the pieces of the jigsaw that would need to be reassembled on
the ground.

First and foremost would be numbers to correctly identify which stars
were being observed as the telescope scanned the heavens, their precise
times of observation, and the data string signifying the detector readings for
each star. Interspersed would be a digital encoding of the satellite's pointing
direction at each moment of time, temperatures at critical places within the
telescope assembly, instrument focus readings, data encryption codes, and
a cascade of other information needed to reconstruct the detailed health of
the instrument.

However, the data stream in its raw digital form was as far removed
from a useful space-age eye on the Universe as a pile of hewn stones was
to the Gothic cathedral at Chartres, or a palette of oils to a canvas of Monet—
crucial, but far from the full story. And they would demand creative insight
and much hard work to transform.

M. Perryman, *The Making of History's Greatest Star Map*, Astronomers' Universe
DOI 10.1007/978-3-642-11602-5_10, © Springer-Verlag Berlin Heidelberg 2010

The preparations for the number crunching tied up the efforts of more than a hundred scientists working across Europe for a decade, such were the intricacies of the data, the complexities of the processing, and the wide range of expertise that needed to be assembled. The analysis was verified by lengthy and tortuous data simulations, processing tests, and endless discussions needed to put the entire system into place. All the work was done not by teams centrally co-located, but by 'virtual' groupings of one or two, sometimes half a dozen, scientists working at a number of different institutes, communicating by technical notes, by mail—and much later by e-mail, and through various meetings.

BEFORE THE SATELLITE could be set in motion to observe anything, however, there was a need for a master catalogue, defining the list of stars to be observed. It would list their positions, as accurately as was known at the time, telling the satellite's pointing system which part of the sky it was scanning, and instructing its detectors which stars were next in line to be observed.

Down to the faint limit of observability of the telescope of around twelfth magnitude, a hundred times fainter than the limit of the human eye, there are some million or more stars in the sky. It was not difficult to figure out how much time could be given to each as the telescope scanned, and how many could therefore be observed over its lifetime. Hipparcos would have time enough to observe only around one hundred thousand, so a careful selection had to be made.

Deciding which out of the million possible were to be observed would itself determine which stars would be in the final catalogue, which would be handed down to future generations with their accurate distance and proper motion measures and, just as important, which not. If a population census was to be carried out for a city of a million people, with time enough to interview just one in ten, the challenge in choosing the sample would be similar: a careful plan and a serious effort would be needed to identify a good distribution according to geographical location, age and gender, profession, income, and ethnic background. So also for the stars.

Cast a glance at the night sky and imagine, not the five thousand or so visible to the naked eye, but one million as seen by a modest telescope, sprinkled around the heavens. The task was to select the one hundred thousand most important. So much was obvious. But which were the most important? Would it be the few known white dwarfs, or the most nearby stars, or those representative of our Galactic disk or its ancient halo? The highest proper motion stars were important, so too were a long list of binary stars and variable stars. Stars with unusual chemical abundances had to be included, along with the oldest subdwarfs. And so the list went on. On top of all, there had to be a nicely uniform distribution of stars across the sky to serve as the celestial reference frame for future generations. For each star included, ten would have to be excluded. For each contented scientist pleased with the wisdom of a certain selection, another might be quite dissatisfied with the myopic choice.

Constructing the starting catalogue was to prove a mammoth task, led by astronomer Catherine Turon of the Paris Observatory in Meudon. In the late 1970s, Turon had discovered the intoxicating grandeur of the project through three colleagues: Jean Delhaye, former director of the Paris Observatory, who had conveyed to her his own curiosity about the structure of the Galaxy; Jean-Claude Pecker who, as their paths crossed in the observatory gardens, had asked her to replace him at an early symposium organised by ESRO to gauge interest in space astrometry; and Jean Kovalevsky, who had urged her to probe the appeal to French astronomers of large numbers of accurate star parallaxes.

Catherine Turon (2005)

Once addicted to the goals, and duly elected to lead the task, she assembled a team of about fifty astronomers ranged across European institutes and observatories to begin the work. Superbly organised, with an encyclopedic knowledge of the stars, she inspired a large team that would work for more than five years to deliver the starting catalogue. Always with a smile, always quick to laugh, she had a passion for the task she had undertaken, and a clear view of the final result that she wanted to achieve. People could not wriggle out of any commitments they had made to her and her team, and excuses for anything deviating from perfection, or the pressing schedule, were not well received. "*Mais non*, that was *not* what we agreed!", but her reprimands were always issued with a winsome smile.

Putting together the satellite's observing list was a balancing act: figuring out scientific priorities of each star, checking the expected performance of a trial catalogue by detailed simulations, assembling the information already known about each object, and setting up new observations using telescopes on ground to fill in missing data needed for the space operations.

There were also a number of very practical problems. Any one star can appear in numerous catalogues, and can therefore be known by many different names. The bright star Procyon, for example, is HD 61421 in the Henry Draper catalogue, GC 10277 from the General Catalogue, FK5 291 from the Fundamental Katalog, LTT 12053 from a high proper motion survey, and so on. You might think that the position of the star would resolve this dilemma, but many catalogues of bright stars often do not list accurate positions; indeed, for many stars, accurate positions had often never been measured. In any case, all stars have a proper motion, and depending on which reference frame and time standard was used—and there were two competing systems at the time—even the position could differ between catalogues. A crucial point was that every star observed would need to have its position known to about one second of arc or better by the time of launch, just to point the satellite's sensitive detector. Getting this accuracy is not too difficult—but it was, in the early 1980s, terribly time consuming.

It had been agreed, during an early phase in the project's development, that the wider professional community would be consulted on their opinions as to which stars should be observed—this was considered a once in a life-time opportunity for science, and the wisest council was, in consequence, sought. I steered through a policy paper which had to be debated and endorsed by the ESA advisory committees before we could open this up to non-European suggestions. The stars observed would form a legacy for decades, and we wanted to make sure that the most important would be observed. This was no time to be parochial, our argument went, and it would be quite inappropriate to restrict scientific opinion exclusively to European scientists. There was the counter view forcefully expressed, that European nations were paying and that, accordingly, it should only be European scientists sowing the ideas and reaping the rewards.

If this should seem small minded, the logic carried force for those who wielded authority: national funding bodies would expect to see a return on their investment, in the form of scientific publications citing their own astronomers' work, not somebody else gaining the plaudits—if you paid a car dealer a fortune, and saw someone else driving off in your Ferrari, you might not be best pleased. Both arguments had substance, and had to be debated. The more altruistic camp held sway, and a world-wide call for observing programmes was issued early in 1982.

A CLOSING DATE OF OCTOBER 1982 was announced. The delivery format was carefully specified. Suggested star lists, in their hundreds or thousands, came pouring in. Scientific authorities from around the world had taken the opportunity as seriously as we had hoped they might. We received lists of the stars most likely to provide a maximum scientific insight into their inner workings. Other lists identified for us objects likely to give the most knowledge about the rotation or structure of our Galaxy as a whole. Lists detailing nearby stars, high velocity stars, rare but important stars like the pulsating Cepheids and RR Lyraes, others mandatory for defining the stellar reference system, important binary systems, bright stars in the Magellanic Clouds, and so forth, all flooded in.

Today, the details could be sent comfortably by e-mail attachment. But twenty five years ago, neither the internet nor e-mail existed. Instead, half the proposals came by post on nine-track magnetic tapes, a bulky storage medium the size of a couple of dinner plates, which could hold a hundred megabytes of data, and which were the state-of-the-art in data storage at the time. The remainder were sent in on the still used, but even by then rapidly-waning medium of punched cards, anachronisms already which had dominated data entry and computer programing for almost half a century[38]. I had commandeered an office for the temporary storage of these tapes and cards, before their onward despatch to the Observatory of Paris in Meudon, which would be the command centre for the next phases. By the time of the proposal deadline, the office was piled high with tapes of different sizes, and punched cards of varying colours. It was an Aladdin's Cave representing humanity's collective knowledge of the stars at that time. I regret not having a photograph to recall the one-time existence of this weighty collection, and of the primitive way that data was stored such a very short time ago.

To pass from disparate lists of suggested stars, to a true master list which could be used to operate a satellite, required a huge amount of work. Redundancies had to be eliminated, obvious omissions rectified. Sky regions too much in demand had to be whittled down. Holes had to be plugged in areas where too few stars had been submitted. Positions had to be checked, and proper motions too. But it was the scientific priorities that would cause the biggest headache.

Adriaan Blaauw, elder statesman in the astronomical world, Director General of the European Southern Observatory between 1970–75 and one-time president of the International Astronomical Union, provided a guiding hand in defining the observing programme.

Adriaan Blaauw (2005)

Following a suggestion by Henk Olthof, the secretary of ESA's Astronomy Working Group, Blaauw was approached, and invited to set up and chair an independent committee to assist. It would be tasked to scrutinise the scientific suggestions, and to assign priorities to the goals laid out. Its brief was to ensure that the starting catalogue observed by the satellite was put together as carefully as possible. In the early 1980s, Olthof took charge of nurturing all new projects entering, or wishing to enter, the privileged ranks of ESA's science programme, and he moved calmly and confidently through the various communities encouraging and facilitating. He took benefit of his compatriot's authority and contacts to rise to the challenge.

Michael Perryman

Selection committee, Paris (April 1987)

Mindful of this unique opportunity to get the list of stars to be observed chosen optimally, Blaauw assembled his own team of fifteen prominent figures in the astronomy world, from institutes around Europe, to contribute their impartial advice. Over the next few years, their task would be to debate the state of knowledge—of nearby stars, stars for the reference frame, stars of specific astrophysical interest, and so on—and adjust the observing programme to reflect the results most demanded from the satellite. Three meetings of the committee over a period of three years were to guide the priorities. Each resulted in a progressive adjustment in the catalogue's contents, all changes to be made in careful dialogue with the leader of the scientific effort, Catherine Turon and her own international team.

OVER SEVERAL YEARS, the team constructing this 'input catalogue' met up for many progress meetings, traveling to one or other of the leading institutes. Astronomers managed to sequester some splendid real estate centuries ago, high points outside major cities, and many have clung onto these superb sites down the years. The observatories of Paris and Nice, Rome and Torino, Heidelberg and Edinburgh are a few that have cornered some of these great locations.

The goal of these meetings was to report on progress, reassess priorities, and debate plans and problems. But they served the additional purpose of bringing team members from different countries into close contact. This fostered a spirit of great collaboration and mutual trust, so essential to the big task building up around us.

One particularly memorable meeting was of the entire consortium, some fifty people, held in the mountain village of Aussois on the edge of the Vanoise National Park in France, in early spring 1985. Turon, supported by her executive team and her international steering committee, drew up plans for a one week conference to get a complete picture of the current state of play of the starting catalogue. Bernard Nicolet who, hailing from Switzerland, had naturally brought his impressively dimensioned Alpine horn along with him, roused us at sunrise each morning with a haunting reveille performed on the slopes outside. At the close of business each day, we could walk up from the Paul Langevin conference centre at the edge of the village into the still snow-covered alpine pastures from which marmots were starting to emerge. Later we might be entertained by impromptu evening concerts from the more musically accomplished. The meeting coincided with the birthday of Pierre Lacroute, the satellite's originator, who was there. A great cake was baked, and there was even some dancing. To talk science for a week in such a location was an inspiration.

On such occasions I was able to meet with some of the senior scientific figures who had dominated ground-based astrometry over the previous decades, including the influential Heidelberg astronomers Wilhelm Gliese (1915–1993), whose name still eponymises the compilation of our knowledge of nearby stars, and Walter Fricke (1915–1988), who led the construction of the state-of-the-art catalogue of ground-based positions and proper motions of stars, the FK5. A small but very select star catalogue,

Meeting at Aussois (1985)

and constructed meticulously following an enormous observing effort over decades, this was the authoritative word on the stellar reference frame before Hipparcos started its own work. Fricke, small in stature, large and jovial in character, had devoted his entire professional life to ground-based astrometry, but became an enthusiastic convert to its future from space. More importantly, he ensured that the formidable expertise of his institute was brought to bear on the catalogue construction: over the next few years his successor as director, Roland Wielen, and catalogue authorities Ulrich Bastian, Roland Hering, Siegfried Röser, and Hans Walter amongst others, ensured that Heidelberg would be as well represented in space astrometry as it had been from the ground. Fricke moved in a world before smoking was proscribed, and at meetings waved vast cigars to affirm his statements. I found him a charming man, and he slipped me some valued advice along the way.

Twenty years hence, we would be looking at thousands of scientific papers making use of the final catalogue. We would be grateful that its unique content had been assembled with such attention and passion. Even before launch, Adriaan Blaauw proclaimed that Hipparcos had served as astronomy's great "vacuum cleaner", its preparations already providing good positions for many stars hitherto unmeasured, cleaning up the confusing plethora of star names, and consistently identifying and labeling the components of binary star systems.

FIVE YEARS AFTER THE WORK on the starting catalogue began, it was completed. The list of stars to be observed was formally delivered to ESA at a ceremony presided over by its Director General, Reimar Lüst, on 11 April 1988 at Noordwijk. Roger–Maurice Bonnet as Director of Science participated, and Pierre Lacroute was guest of honour. Catherine Turon handed over a magnetic tape to Reimar Lüst in a symbolic gesture of an important milestone—the magical list of stars that had cost so much effort to prepare.

VIPs visiting the satellite, 11 April 1988

After the ceremony, Hamid Hassan and I led Lüst and Bonnet, Pierre Lacroute and Catherine Turon, Erik Høg and Jean Kovalevsky along the warren of corridors, through the security barriers, and into the integration vault where, coated and masked, we could see the satellite being assembled under 'clean room' conditions. Wires hung everywhere. Motors hummed, and lights flashed as checks progressed.

For the last time, we could gaze upon this bizarre construction of glass and metal, and reflect on the complex combination of circumstances which had led to its creation.

GETTING READY FOR THE DATA PROCESSING was an even larger challenge. There were two distinct problems: how in principle to formulate the precise mathematical equations to extract the star positions from the data, and how in practice to crunch through the huge mountain of numbers that would soon be falling to Earth.

Even if, somehow, the satellite could have been designed, built and launched without recourse to computers, the calculations to go from the raw data to the final catalogue could simply not have been done by hand. Yet even this statement takes no account of the human ingenuity, and sheer manpower, that might have been thrown at the problem if there had been no other option.

Until about four hundred years ago, calculations in astronomy as elsewhere were performed by hand. Navigation as one of its main applications involved spherical trigonometry, which relates angles and lengths of circular arcs on the sky. The earliest of calculations used trigonometric tables, and laborious long multiplication, and there was no choice other than to knuckle down and work through pages of sums with pen and ink.

The last years of the sixteenth century saw the introduction of a quick and approximate way of multiplying two numbers called prosthaphaeresis. This unwieldy word is a contraction of the Greek for addition and subtraction. The method was used by Tycho Brahe, but remained in favour for no more than a couple of decades before being swept aside by the more powerful method of logarithms.

Introduced by John Napier in 1614, and embraced by Kepler and others, logarithms remained in use for more than three hundred years until the 1950s and beyond. It was used extensively, and taught in schools, certainly through the 1970s. [A short tutorial for those who do not know and may be interested: if you want to multiply 1000 by 10 000, you can write the former in powers of ten as 10^3, and the latter as 10^4. Instead of multiplying the numbers you can add the exponents, here $3 + 4 = 7$, and obtain the answer 10^7 or ten million. Multiplication has been replaced by the simpler task of addition.] Logarithm tables paved the way for the multiplication of any numbers, not just powers of ten: to multiply two numbers, look up their tabulated logarithms, add them, and consult the anti-logarithm to find the answer. For those tempted to chuckle and reach for their calculator, how many would know how to construct *its* innards?

Tables were created manually, and compiled in book form. Already in 1617, Henry Briggs had worked out the logarithms of all whole numbers below 1000 to eight decimal places, an early example of many such heroic efforts. The Paris Observatory contains the single copy of the monumental *Tables du Cadastre* of 1792, with the logarithms of all numbers up to 100 000 to nineteen places, and of the numbers upwards to 200 000 to twenty-four. The scale of the mind-numbing effort required almost defies comprehension.

Computers entered the professional arena in around 1950, and moved swiftly from vast power-hungry racks that could handle only the simplest of sums, to the lightning speeds of billions of calculations each second done on a personal computer today. In an astonishing bridge across history, Danish astrometrist Erik Høg, at the forefront of the Hipparcos and Gaia satellite designs into the early 2000s, himself recollects a university lecture by Peter Naur in Copenhagen in 1952 recounting his first exposure to the Cambridge-built electronic computer EDSAC. "The room was overflowing", Høg recounts, "I was sitting with others on the floor, and even Niels Bohr had come to listen and to wonder at the fantastic punched tapes with rows of holes for numbers." Their power and performance has accelerated relentlessly since then.

EDSAC I (c. 1948)

Computer Laboratory, Cambridge University

AS THE SATELLITE WAS BEING designed and built in the early 1980s, we planned that the data would be sent down from the satellite, from around 1990, at what was, at the time, a worryingly large rate of twenty four kilobits a second. Time marches inexorably on, and this wouldn't cause much of a raised eyebrow today. But when the project was being sold to the ESA advisory committees in 1979, and still when the data analysis system was being designed in the early years of the 1980s, it represented a daunting task. More crucially, there was no direct way of solving the gigantic system of mathematical equations posed by the data stream. Even into the early 1990s, neither computer memory storage, nor brute force computational power, could possibly crack the data processing demands.

Enter the work of Lennart Lindegren, a towering figure in various design and analysis aspects of the Hipparcos satellite concept, and who we met already in the formulation of the beam-combining mirror. In the late 1970s Lindegren, a quietly unassuming Swedish astronomer from the University of Lund, had been introduced to the problem of the data analysis by Erik Høg. Høg, whose own insights into astrometric instrument design were arguably unparalleled in the last three decades of the twentieth century, has often ribbed self-deprecatingly that one of his greatest contributions to astrometry was finding Lennart Lindegren, and enticing him to join the growing ranks of Hipparcos scientists. Whatever one achieves in science, encouraging a new fertile mind to grasp the intellectual baton and head off towards further unimagined advances is indeed exciting.

Within a month of being introduced to the problem in September 1976, Lindegren had put forward a clever and elegant short-cut to the rigorous intricacies of the data analysis. He split the problem into three sequential steps, an insight which thereafter dominated the data processing throughout the project's life and which ensured its eventual tractability: first, estimate the separation of all pairs of stars observed along a single full scan of the celestial sphere; next, solve for star positions in just one dimension for each of these circles; finally, interlock all these circles, estimate their starting points, and insert these back to find the position, motion, and distance of each star.

Lennart Lindegren (2009)

Like many first-rate ideas, it was rather obvious once described and mathematically elucidated. It represented the Eureka moment for the number-crunching on ground, and the turning point in proving the feasibility of the huge computational analysis task which lay ahead. It set in motion the work of a large group of astronomers and programmers who would or-ganise themselves to prepare the software, run simulated data through the entire system, and to ensure, by meticulous preparation, innumerable meetings, countless checks and cross-checks, and careful documentation, that the software would do what it was set up to do. Important, also, was to assess and negotiate with the operations team and the satellite's industrial architects, that the information sent down from the satellite would follow a strict protocol that could be digested in the data treatment on ground. Information for each star had to be uniquely recognisable, the signals correctly encoded, and the satellite spinning motion precisely represented. No crucial information could be missing.

I will not touch on the numerical techniques, and the vast sweep of complexities that had to be accounted for in the data processing chain, except for a brief foray into the corrections needed to account for two of Nature's more subtle massaging of star positions that we encountered in relating the historical development of astrometry: the twin puzzles of stellar aberration and general relativity.

Aberration is the tiny shift in a star's position which results from the speed of a moving observer being added to the speed of light, which James Bradley had stumbled upon a quarter of a millennium earlier. To account for the effect, both the satellite's position and velocity had to be known with high accuracy at each and every moment, along with the position and velocity of the Earth itself. Satellite distance ranging from the various ground stations provided the satellite position to an accuracy of one kilometer at all times as it raced around the Earth, and its velocity to better than a few centimeters a second. Experts in the dynamics of the solar system at the Bureau des Longitudes in Paris, and at the Jet Propulsion Laboratory in Pasadena California, provided us with the careful records of the position and motion through space of the Earth itself. Thus we would have the knowledge to correct for the aberration of starlight.

Bending along the path of starlight in the vicinity of the Sun due to the shifts described by general relativity had to be similarly accounted for. When two of the most highly renowned astronomers of their day, Arthur Eddington and Frank Dyson, first measured light bending due to the massive Sun at the time of the solar eclipse in 1919, the effect was only just detected with the instrumentation then available: it amounted to a little more than one second of arc for light just grazing the edge of the solar limb where this most tiny and subtle of effects was at its maximum. The phenomenon, however, plays out everywhere on the sky and at all times, albeit at still more diminuitive amounts the further from the Sun the starlight passes. Tiny though the effect was in the context of observations made almost one century ago, it assumed enormous significance for the Hipparcos space measurements, an effect of importance over the entire celestial sphere. As the satellite raced around the Earth, and as the Earth moved around the Sun over its three years of life, all stars across the entire sky appeared to gyrate in a complex pulsating quiver, as the Sun's colossal mass distorted the underlying fabric of space and time.

Complex though this pattern of apparent motions was, the details of the distortion pattern is described by just one single number: the degree to which space–time truly agrees with the prescripts of general relativity. The extraordinarily delicate Hipparcos measurement of light bending over the entire celestial sphere duly provided perfect confirmation of general relativity. Of even greater importance for the majority, who had little doubt that the theory would be vindicated even at this level, it provided a reassuring check on the validity of the first astrometric data obtained from space.

IN A SOFTWARE SYSTEM BUILT UP by dozens of scientists over many years, the potential for errors, perhaps deeply hidden within hundreds of thousands of lines of program code, is significant. In such a project, at the boundaries of scientific knowledge and with no external checks possible, errors could lay undiscovered—subtle perhaps, but potentially catastrophic for scientific studies that would be carried out on the final catalogue.

The solution adopted was to duplicate the entire data processing chain, starting with the parallel ingestion of the thousands of magnetic tapes of satellite data, and thereafter all the way on down through the three-step analysis. Of course, this approach involved additional expense and duplicated expertise. And it only made sense if ways and means of comparing and correcting any errors could be put into place.

Erik Høg's 'NDAC' consortium (Greenwich, 1984)

Fortunately there were enough institutes in Europe willing to pool their knowledge and resources, and two separate data processing groups were accordingly set up, broadly demarcated into northern and southern European efforts. The groups would develop their entire processing chains independently, and in due course, throughout the satellite harvest, lasting from soon after launch in 1989, through to the end of operations in 1993, and on to the final catalogue completion in 1996, process the satellite data independently.

Erik Høg (1982)

Jean Kovalevsky (2005)

The rules of engagement were soon agreed: methodologies could be discussed, but it was accepted from the outset that all the software code would be developed independently. A common vision of the best possible end product ensured that the entire undertaking was carried out in a spirit of friendly rivalry. Through the mediating forum of the science advisory team, the progress of each group was presented, challenged, and adjusted to take account of new studies and evolving insights.

It is difficult to overemphasise the vast effort that was needed to design, write, and put into place the computer software for the entire data processing chain. Together, around a hundred scientists pooled their expertise over more than a decade, writing, testing, and retesting enormous tranches of code which would step, automatically, through the complex data after its arrival on the ground.

Erik Høg from Copenhagen set up and led the first group, enrolling a small and highly competent team focused in northern Europe, comprising astronomers from the Royal Greenwich Observatory who would take care of step one, a small group centred on the Copenhagen University Observatory for step two, and Lennart Lindegren at Lund Observatory in Sweden for step three. Each institute would be responsible for software coding before launch, as well as the analysis of the satellite data for their assigned step. After launch, data was despatched from the operations centre in Germany to the RGO, and thereafter passed to the next in the processing chain using magnetic tapes sent by normal postal mail. The Greenwich team started work at its imposing premises at Herstmonceux Castle in Sussex, moving shortly before launch to its new home in Cambridge.

Jean Kovalevsky from Grasse led the second, a substantially larger team with a different processing philosophy. He assembled groups from France, Germany, Italy and The Netherlands who would develop the various pieces of the entire processing chain. Thereafter, their software segments were delivered, before launch, to a central institute of the French space agency,

CNES, in Toulouse. There, the software was tested and validated, then all the individual pieces of code were cemented together into a single end-to-end processing system. Their copy of the satellite data was delivered to this single location, and all of the number crunching was handled there. The larger team required more skill to manage, a challenge Kovalevsky rose to with a committed technical executive board, and an international steering committee. Step-by-step this huge software system fell into place: the basic star observations and satellite attitude treated by software from Torino, instrument calibration from Utrecht, the celestial circle scans by geodesists from Delft, sphere mapping from Torino and Grasse, the star motions and distances from the Astronomisches-Rechen Institut in Heidelberg, double stars from a team distributed in Asiago, Bari, Frascati, Grasse, and Heidelberg, and the operational system itself in Toulouse.

Central to its smooth execution, a vast multi-institute effort spearheaded by Michel Froeschlé generated huge swathes of simulated data to test the system before launch—long stretches of the raw data expected from the satellite, others simulating the scanning circles, the satellite's attitude, and many other pieces of the entire data jigsaw.

For both processing teams after launch, and running alongside ESA's own massive headache of redesigning the satellite operations to accommodate the unintended orbit, almost a year of extra effort was needed to redesign the software on the ground, to handle the new orbit and the very different forces affecting the satellite's spin. Then slowly, as the real satellite data processing got under way, the massive celestial jigsaw was progressively assembled: star images crossing the grid were tracked, the attitude reconstructed, circles around the sky assembled and, slowly but surely, the positions and distances and space motions of all the stars started to appear.

It would have been folly to let the two systems run to completion before comparing the results, yielding two separate catalogues which turned out to be slightly different. The cause of any difference could still lay buried within the mountains of software code, all but impossible to discover. Instead, comparison points were identified, and differences discussed and resolved by expert groups before moving on to the next step.

Any wrinkles were smoothed out progressively, and the catalogues slowly proceeded towards completion. Some steps had to be debugged, fixed and re-run, using outputs of some of the later steps as input to a repeated analysis of one of the earlier steps. For Erik Høg's geographically distributed consortium of the early 1990s, magnetic tapes shuttled back and forth by mail between the UK, Danish, and Swedish institutes; a logistical complication which would have been circumvented by the data highways of only a few years later.

As more and more observations were collected, and better positions calculated, other tasks intensified. One group led by François Mignard from France, with Staffan Söderhjelm and Lennart Lindegren from Sweden, gathered all of the individual position measurements, and reconstructed tens of thousands of binary star motions and orbits. Dafydd Wyn Evans in Cambridge led a team which carefully calibrated the millions of magnitude measurements, passing them on to Floor van Leeuwen in Cambridge, and Michel Grenon and Laurent Eyer in Geneva, who assembled from them the beautiful light curves of the thousands of variable stars detected by the satellite.

François Mignard (2001)

Amateur astronomers, their work coordinated by Janet Mattei of the Amateur Association of Variable Star Observers in Harvard, assisted with critical observations of irregular variable stars made from the ground.

ONE FURTHER PART of the data processing on the ground needs to be mentioned: a separate data stream which led to the separate catalogue, titled the Tycho catalogue by Erik Høg in recognition of his illustrious compatriot Tycho Brahe four centuries before. Høg, who occupied a key role in the selection, design and execution of the mission, traces his own involvement in astrometry through his teachers in Copenhagen, Peter Naur and influential astrophysicist Bengt Strömgren (1908–87).

In April 1981, already well into the satellite design phase when significant modifications would normally have been strictly rejected out of hand for reasons of risk, and for the increased cost that they could incur, Erik Høg had pointed out that a rich seam of satellite data was lying untapped. He realised that the signals from the satellite attitude detectors contained a staggering quantity of star positions that were, quite simply, not being sent to the ground. But once pointed out the harvest was obvious: positions for the million or more stars not being observed by its main detectors.

We heard Høg's ideas for the first time at a science team meeting in 1981. There was silence while we digested the treasure ahead, the work needed to recover it, and how close we had come to missing it. We knew that to pursue it would be difficult because of the timing, and would involve much extra effort. But it was equally obvious that the opportunity could not be ignored. Working quickly to optimise the system, and with a little urgent lobbying, the small amount of additional funding needed was made available. A filter and additional detectors were added, an extra telemetry channel set up to send the data to ground, and a new data processing team put in place to handle the data stream.

Data tapes leaving Tübingen (1992)

After launch, in 1990, Erik Høg passed his leadership of the main data processing chain to Lennart Lindegren, and applied himself full time to the growing task of running the Tycho team. This now embraced astronomers from most of the ESA member states, but the lion's share of the work was done in Denmark, and assisted by a team of ten young and capable astronomers from Heidelberg, Tübingen, Strasbourg and elsewhere in Europe. The next few years would show how fortunate astronomy had been that this additional source of star positions had been tapped.

THE WEEKS AND MONTHS AFTER LAUNCH were an anxious and impatient wait for the operations to start up, and to run more smoothly. But after the first data from the satellite arrived on the ground, another blow appeared to fall upon us: the first pieces of data did not snap into place along the celestial circles as expected. The observations made no sense. For a few days we were, not for the first time, deeply concerned. Had something fundamental been overlooked?

Hans Schrijver (2007)

Hans Schrijver, working in Kovalevsky's team, had set up a one-man first-look data quality verification facility at the Space Research Laboratory in Utrecht, scrutinising a few hours of satellite data each week to quickly verify that they were up to scratch. Instrumental tweaks could be relayed back up to the satellite, to adjust settings like the telescope focus if not. Schrijver quickly found the answer. The modulating grid at the telescope's focus was ever so slightly rotated compared to its position measured on the ground. Corrected for this tiny rotation, the data fell immediately and beautifully into place.

As the data started to flow in, the relentless computations began. The following months and years were to prove a roller coaster ride—high points as the data mountain piled up and the catalogue started to fall into place, low points as the radiation damage took its unpredictable and debilitating tolls: killing off the crucial gyroscopes, or wiping out, one by one, the instrument heaters, so important for keeping the telescope optics stable. Equally unpredictably, and most painfully, from time to time some hours or days of data were lost as solar flares erupted and swamped the data taking. Under these anxious circumstances, the satellite performed an emergency autonomous slew back to a safe Sun-pointing configuration, and the operations team would be called out, at whatever time of the day or night, to restart the 'grand slam' manouevre back to the proper scanning motion.

As the weeks went by we had the first of the great circles, and a progressively better understanding of precisely how the satellite was turning in space and scanning the sky. The results of the first circle of data, the first few hours, were exciting indeed. The positions of the stars observed along the scan really seemed to have been improved. After six months, the first full sky solution gave us the first catalogue of star positions. Imperfect when compared to the final results which were eventually to be expected, they were vastly better than any measurements that had ever been achieved on the ground.

Yves Réquième at the Observatory of Bordeaux used the new positions to see whether past meridian circle observations made on the ground would be improved by the new star positions culled from space. They were, and it was another exciting moment. It meant that the observations, and the analysis, were proceeding to plan. It meant that decades of star positions made from telescopes on ground could, in due time, be reprocessed and improved using the space-based positions. Things were quickly and most satisfactorily falling into place.

After a year of data, a greatly improved solution for all stars together gave the first hints of the stars' proper motions, vastly improving on the ground-based values. We crossed our fingers that the satellite would continue to work.

After eighteen months the first parallaxes started to drop out. We had our first star distances from space. Hipparcos was delivering everything that it had

Jean Kovalevsky's 'FAST' consortium (Bari, 1986)

promised. Beyond this moment, the solutions just got progressively better and better. Binary stars could be solved for. Variable stars were recognisable in the star intensities. The end was now in sight.

The final data from the dying satellite was sent to ground and fed into the processing chain in early 1993. Already badly crippled with inert gyroscopes and payload heaters, contact with the satellite failed irretrievably on 15 August 1993. Intensive care was discontinued. It was a poignant moment to accept that the satellite was now nothing more than an inert and useless piece of Earth-orbiting debris. But all the data that was needed was now in the bag. We could be confident that Hipparcos would, in due course, be a complete success. So it was a triumphal moment as well.

From then until mid-1996 when the catalogues were completed, the solutions were iterated, adjusted, compared, and improved. A line was drawn under the processing in mid-1996—further improvements could be made, but after three years we were into a regime of diminishing returns. Computers had whirred across Europe for more than six years, digesting a billion star separations in total, along with some three hundred million gyroscope readings, interlocking them all to create history's greatest star map.

THE WIDER ASTRONOMY COMMUNITY WAS STARTING to clamour for access to the data. The demands for early access were understandable, a testament to the mission's growing interest, but from some quarters, they were unreasonably insistent and somewhat unwelcome. In setting up the schedule of the data processing in the years before launch, ESA and the scientific teams had signed a detailed Memorandum of Understanding. Amongst the various conditions underpinning the collaboration was the target goal of finalising the catalogue three years after the end of the satellite operations, with the agreement to go public with the results one year after that. This one year fallow period would allow for documenting the results, preparation of the printed volumes, and one year of proprietary access rights for the proposers who, more than a decade before, had submitted their ideas to the observing programme selection committee and, by so doing, had ensured the lasting value of the stars selected for observation, and traded this for guaranteed early access to the final mission results before their wider public release.

Although the agreement had been signed off for all to see, and duly approved by the advisory committees, it didn't diminish a vocal minority who wanted to know why the data release was taking so long. The consequences for acquiescing to these demands for early access could easily be predicted: satisfaction in having their hands on preliminary data would quickly be replaced by widespread confusion and criticism as different incarnations of the solutions became available. The science team dug in its heels and refused to be rushed in its delicate task. A painful episode at the time, I was assured by wise counsel that such unfounded criticism of the project's progress would quickly be forgotten, which indeed it was.

THAT AGREEMENTS ARE NOT ALWAYS WORTH the paper they're written on is a frequent caution, but one that can still take us by surprise. It was with some consternation that we learnt, one day early in 1991, with the data processing in full swing at last, that the UK Science and Engineering Research Council was intending to terminate its funding for the Hipparcos team at the Royal Greenwich Observatory. At times like this one can question one's hearing if not one's sanity, or cast a quick glance at the calendar to see whether April Fool's day has crept around again unannounced. Their reasoning was that the processing was being duplicated by two teams (which was true), that one was therefore redundant (which it wasn't), and the research council was short of money, £80 000 to be precise, now needed for other projects. That the total amount being contested was comparable to the weekly wage of a premier league football player puts the debates that often occur in science funding into some context. The total cost of the Hipparcos project to the UK over twenty years, for example, was somewhat less than the cost of one mile of a dual three-lane motorway, estimated by the Highway Agency in 2005 as £25 million per mile. Big science *is* expensive, and society as a whole has to judge its relative merit.

But this sudden U-turn in funding support flew in the face of a written agreement in place between ESA and the scientific teams. It was a unilateral decision that would pull the rug out from under the feet of the UK–Denmark–Sweden consortium, and would terminate at a stroke the project's carefully planned and ongoing capability to rigorously cross-check the results. A wholly unexpected development, the crisis was passed quickly up through the ESA advisory committee structure like a hot

Space partners fall out over British cuts to star map project

Robin McKie and Nicholas Booth

A BITTER conflict has broken out between Britain and its European space partners over plans to halt UK funding for a unique star-mapping probe that is circling the Earth.

cussions beyond the country's involvement in the project. Analysis of three years' data is required to improve our knowledge of star positions. Any disruption would curtail the project indefinitely.

The threatened cut infuriated Esa delegates when they were

© Guardian News & Media Ltd 1991

The Observer, 24 March 1991
(Robin McKie and Nicholas Booth)

potato, landing in the lap of the Science Programme Committee where various delegations whose hands moved the levers of power voiced their dismay with the UK's position[39].

Arnold Wolfendale, chair of the SERC's astronomy and planetary science board, recently appointed as Astronomer Royal, and UK delegate to the Science Programme Committee, was caught squarely if unfairly in the midst of the furore, and was forced to hold out for an independent study to evaluate the implications of its actions. "I know it makes no sense to withdraw from Hipparcos now," he said[40], "but then it makes no sense for the Government to cut back on the country's science budget in the massive way that it has."

The debacle tied up the Hipparcos teams in some weeks of counter response and renewed justifications, and bled money and frustration in the
process. Fortunately, within three months the UK funding authority had
taken objective council, and reinstated its support. It was a blemish on an
otherwise remarkably problem-free decade and a half of unqualified support by the member states of ESA, at all levels, to the Hipparcos programme,
although it left anxious questions about the reliability of international scientific collaborations which would take long to subside.

High office carries the burden of such difficult decisions. And while
we had crossed swords in the execution of our respective duties, when
Hipparcos was eventually switched off, in August 1993, Arnold Wolfendale
generously penned a letter to me with his congratulations on its completion:
"The dedication and skill shown by your colleagues and yourself in rescuing
this mission and going on to produce such superb results will go down in the
history of space science."

A MORE JUSTIFIED FUNDING contest had to be faced a year later. Operations
were costing more than originally budgeted due to the extra ground stations,
and to the larger operations team needed to support it around the clock. The
approved funding would be exhausted in January 1993, and I had to set off on
the road again—the agreement of the advisory committees had to be sought,
and the largesse of the Science Programme Committee appealed to.

I had the pleasant task of explaining progress to date and the modest
sum requested, especially so because I was bidden to the beautiful island of
Capri where, shimmering in spring sunshine, and at the invitation of the Italian delegation, the Astronomy Working Group and Space Science Advisory
Committee were meeting in May 1992. Support of the former was briskly
affirmed. Approval of the latter took some unexpected twists and turns as
the relative priorities of Hipparcos and the International Ultraviolet Explorer
satellite, also still in orbit, were aired. Director of Science Roger–Maurice
Bonnet expressed his unqualified support for Hipparcos, while Belgian Jean–
Pierre Swings, as chair of the Astronomy Working Group, and Italian Cesare
Barbieri declined to debate the issue given that case was compelling and the
money at stake so small. I had laid out the scientific justification and could
do no more. Leaving the delegates in animated discussion but with the outcome reasonably secure, I enjoyed the remainder of the afternoon strolling
the island lanes towards Anacapri, perfumed with the intricate scents of the
sparkling Amalfi coast.

The money was not in the bag, however, until the Science Programme
Committee declared it to be so. Accordingly, a month later, I was commanded to ESA's headquarters in Paris, and to the final grand hearing held in

the hushed confines of the deeply-carpeted Council Room, my request translated into a babel of languages for the pan-European delegations. Guided by the chair, no-nonsense Dutchman Johan Bleeker, director general of the Space Research Organisation in The Netherlands, the necessary sum was eventually nodded through in a consensus of eager approbation.

THE FINAL CATALOGUES, done and dusted, were carefully documented and prepared for publication, a task taking a further year. While the four years of operations of the satellite had been particularly stressful, pulling together all the pieces of the final catalogue for publication was relentlessly hectic, and occupied many of us, fully, for most of 1996 and early 1997. Different groups had different ideas of what data had to be included, and how it should be presented, while the technical and publication details occupied several of us, without respite, for months. Amongst the dozen or so shouldering the leading responsibility, fourteen hour working days, week after week, were not uncommon—despite overtime pay and bonuses not being part of the scientific reward package.

Andrew Murray from the Royal Greenwich Observatory and Frédéric Arenou from the Paris Observatory pulled off a remarkable exercise of figuring out how to blend the two parallel results into a unique catalogue for the world's astronomers to use. Recently retired from the Royal Greenwich Observatory, Murray had probably witnessed a more profound change in positional astronomy over his professional life than anyone else in the team. His early career at Greenwich had included his part in the ancient routine of dropping the time ball, that large and prominent sphere that fell from the observatory tower at a predetermined time to enable mariners to check their sea-going chronometers for navigation, and which had remained in use until the introduction of radio time signals in 1924.

Andrew Murray (1997)

Frédéric Arenou (2007)

A scientific committee was set up to document the results, and to come up with an optimised catalogue layout—a tricky task because each star had, associated with it, around a hundred separate pieces of information that needed to be conveyed. Much thought went into the format for publication. The scientific results themselves were stunning, and we wanted the published catalogue to reflect the innate beauty and elegance of the information that it held. I kept track of the ideas and suggestions, and we went through more than twenty tweaks to get the layout which said it all, in the way it had to be said.

Central to the publication was another one-man virtuoso performance pulled off by Utrecht scientist Hans Schrijver. He set up a special data base, in the premises of the Dutch Space Research Organisation in Utrecht, into which he fed all of the results coming from the various parts of the processing teams—the main catalogues as well as magnitudes, binary star information, notes, references and the whole host of other sparkling gems that would all be published in the seventeen volume treasure chest—duly validated, and thence output to a master printed form to be subsequently published.

Despite the heavy workload, events sometimes occurred which could nevertheless make us smile. On one occasion, I had been working through a Saturday and, before quitting, had made a suggestion to include a table of nearby stars. I raised it in an email to Lennart Lindegren in Sweden, and copied the details to Hans Schrijver in Utrecht and François Mignard in Nice, inviting their own thoughts. Over the next twenty four hours Lindegren, who had also been working around the clock, responded with a specific suggestion, and Mignard, also at his computer terminal that weekend, had made a further improvement, which Lindegren and I were able to agree to by late Sunday night. The problem was solved. But as Hans Schriver admonished us when arriving at work early on Monday morning: "Well, well, one cannot afford to be out for the weekend on this project; even during such a brief absence, very good ideas can be suggested, and improved, and accepted!"

I PERSUADED BRUCE BATTRICK, head of the publications division in ESA, to throw his team, his experience, and his resources behind the publication. We had worked together frequently in the past and, being a physicist by training and a motivated advocate of outreach, he needed little persuasion. His colleague, Henk Wapstra, took control of the overall logistics. The first four volumes documenting the undertaking were put onto film in ESA, the next nine volumes came from Hans Schrijver via the Imprimerie Louis–Jean from Gap in France, and the three-volume Millennium Star Atlas originals were created by Sky Publishing Corporation in Cambridge, USA, under the remarkable leadership of Roger Sinnott.

As printing progressed at the firm of Spiegelenberg in Zoetermeer in The Netherlands, dozens of wooden pallets teetered high with a thousand copies of the seventeen volumes awaiting binding, strewn around the warehouse with what seemed to me chaotic abandon. Computer scientist William O'Mullane joined me in ESA to transfer the data to the new and still somewhat mysterious medium of CDs, mastered and duplicated by Swets & Zeitlinger from Lisse, while Karen O'Flaherty set about unlocking the new challenges of posting the data on the www. Technology had marched alongside us, providing powerful and unexpected means of circulating the catalogues around the world.

Catalogue printing (1997)

Three decades after the idea of a space astrometry mission had been mooted, the tangible banquet of results was laid out before us, ready to be tasted and digested.

THE PRINTED VOLUMES may seem a curious extravagance, although with only a thousand or so sets produced, it was a minor financial perturbation to the project as whole.

Five hefty volumes together represent the positions, motions, and distances of one hundred thousand stars—volumes superficially resembling a densely typeset telephone directory. Others give light curves for variable stars, details of binary stars, and a whole host of more arcane data of interest to the specialist. Mindful about the fragility of consigning unique science to evolving storage media and faceless hardware servers, the decision to print the key results of the mission was taken with an eye on longevity. Acid-free paper, and a fiendishly clever computer 'check-sum' hidden on each page, should mean that the pages can be scanned in, reliably, at any time in the future. In a step akin to making provision for the demise of our own computer civilisation, some third millennium re-enactment of the sacking of the great Library at Alexandria, the Hipparcos catalogue might still survive for future generations to use. Overly paranoid, possibly, but as ESA's chief scientist for the project, it was a decision that I had no hesitation in signing off.

At the same time as the great Hipparcos catalogue was published in 1997, the Tycho catalogue of just over one million stars was also finalised. Of lower accuracy than the main catalogue, it was for many more stars, and had the great merit of including the colour of each.

The catalogues, at last, were done.

Number HIP 1	2	Descriptor: epoch J1991.25 RA h m s 3	Dec ±° ′ ″ 4	V mag 5	6	7	Position: epoch J1991.25 α (ICRS) deg 8	δ deg 9	10	Par. π mas 11	Proper Motion μα* mas/yr 12	μδ mas/yr 13	Standard Errors α* mas 14	δ mas 15	π mas 16	μα* mas/yr 17	μδ mas/yr 18	Astrometric Correlations (%) δ/α* 19	π/α* 20	π/δ 21	μα*/α* 22	μα*/δ 23	μα*/π 24	μδ/α* 25	μδ/δ 26	μδ/π 27	μδ/μα* 28	Soln F1 % 29	F2 30
32301		06 44 37.45	-38 26 12.3	8.80		H	101.156 050 02	-38.436 752 52		-0.45	5.30	-0.21	0.75	0.76	0.91	0.97	1.19	-3	+17	-9	+12	+0	-4	+2	-13	+13	-3	0	-2.25
32302		06 44 38.52	+19 59 15.4	8.68		G	101.160 515 05	+19.987 620 00		2.28	-0.89	-2.08	1.17	0.69	1.34	1.45	1.02	+12	+27	-32	-12	-29	+31	-3	-63	+40	+41	6	3.43
32303		06 44 39.94	-14 43 42.4	8.83		H	101.166 427 36	-14.728 444 16		8.48	-18.80	-31.95	0.94	0.82	1.07	1.26	1.07	-4	+32	+2	-25	+2	-8	+4	-30	+10	-36	0	-1.53
32304		06 44 41.19	+38 46 41.4	9.56		H	101.171 613 91	+38.778 176 56		5.72	-1.90	-24.34	1.33	0.85	1.47	1.69	1.08	+7	+29	-1	-17	-19	+9	-16	-53	-4	+15	0	-2.67
32305		06 44 42.19	+59 38 26.0	9.69		G	101.175 777 63	+59.640 564 09		2.61	-0.75	-5.55	1.22	1.13	1.58	1.60	1.15	-2	+25	-26	+6	+29	+8	+28	-4	+5	+23	0	-0.40
*32306		06 44 42.94	+26 07 48.7	9.20		H	101.178 897 28	+26.130 197 74		2.23	5.54	-22.38	1.55	0.84	1.51	2.08	1.30	-20	+16	+8	+23	+20	+16	+45	-53	-6	+18	0	0.08
*32307		06 44 42.89	-57 26 25.6	8.55		G	101.178 694 46	-57.440 432 20		10.93	-52.24	56.74	1.21	1.18	1.29	1.37	1.27	-6	+12	+3	+4	-6	-6	-11	-21	-3	-28	0	
32308		06 44 42.97	+14 54 37.4	10.73		H	101.179 036 65	+14.910 388 44		18.07	12.45	-168.02	2.21	1.27	2.26	2.40	1.72	-10	+43	-12	-17	+4	-7	+11	-64	+5	+7	0	-0.06
32309		06 44 44.22	-61 13 27.3	6.95		H	101.184 250 53	-61.224 248 79		3.31	-8.59	8.03	0.55	0.49	0.54	0.61	0.58	-2	+11	+10	+8	-7	-7	-8	+5	-5	-16	0	-0.79
32310		06 44 44.35	-19 03 10.7	7.32		H	101.184 780 68	-19.052 981 12		3.00	-6.54	1.13	0.42	0.48	0.74	0.51	0.60	+5	-2	-2	+15	+11	-10	+7	-5	-1	-13	0	-0.87
32311		06 44 45.46	+28 58 15.6	5.42		H	101.189 436 33	+28.970 989 12		5.72	-8.75	-23.65	0.79	0.43	0.84	0.96	0.64	+19	+33	-7	+3	-13	+2	-4	-42	+22	+24	0	0.31
32312	H	06 44 45.63	-04 23 36.8	8.55	1	G	101.190 105 63	-04.393 555 80	A	4.10	-5.45	-1.18	2.03	2.00	1.92	1.62	1.42	-25	+20	-11	-27	+1	-32	-8	-17	-26	+14	0	1.33
32313		06 44 45.92	+71 53 20.3	10.92		H	101.191 327 78	+71.888 964 31		44.43	-118.98	-549.44	2.67	3.56	4.88	2.94	4.18	+4	+11	+2	-5	+3	+9	+3	-13	+6	+7	1	
32314	H	06 44 46.27	-52 25 25.0	8.29		H	101.192 792 15	-52.423 618 21	A	4.98	-6.12	12.63	1.24	1.23	1.18	1.25	1.33	-6	-4	-1	-21	-9	+19	-9	-18	-1	-22	1	-0.36
32315		06 44 46.60	+44 30 20.0	6.64		H	101.194 150 53	+44.505 542 19		2.53	1.32	-16.23	0.68	0.51	0.79	0.89	0.61	-9	+11	-2	+2	-16	+8	-14	-25	+10	-11	0	1.72
32316		06 44 46.66	-09 23 38.7	7.79		H	101.194 436 11	-09.394 078 52		4.88	11.76	-4.50	0.86	0.68	0.96	0.93	0.80	-26	+36	-14	-39	+8	-4	+12	-50	+7	+0	0	-0.04
32317		06 44 48.95	-28 23 39.8	9.97		H	101.203 976 55	-28.394 400 35		2.64	-13.78	6.92	0.94	1.14	1.56	1.15	1.34	-6	+8	+14	+6	-8	+17	-10	-3	-4	-4	0	2.81
32318		06 44 49.57	+03 40 47.7	7.63		H	101.206 552 04	+03.679 903 28		0.65	-1.82	-0.29	0.89	0.54	1.05	0.87	0.68	-5	+45	-4	-9	-6	-24	-11	-45	-11	+1	0	0.52
32319		06 44 50.94	-18 34 13.6	7.07		H	101.212 241 21	-18.570 445 70		11.21	-16.00	-18.46	0.43	0.48	0.79	0.52	0.59	+1	+5	-13	+22	+24	-9	+20	-12	+2	-12	0	-1.39
32320	H	06 44 51.63	+59 26 57.3	7.07		H	101.215 130 15	+59.449 252 74	A	9.43	-2.67	11.42	1.10	0.89	1.24	1.30	0.81	-22	+51	-30	+12	+5	+29	+16	-19	+13	-1	2	1.54
32321		06 44 51.80	+49 20 37.1	7.99		G	101.215 818 18	+49.343 650 45		7.42	-41.40	-14.85	0.78	0.64	1.04	1.02	0.73	-13	+31	-37	-1	-18	+15	+1	-33	+25	+26	0	1.58
32322		06 44 51.96	-27 20 32.6	6.43		H	101.216 483 45	-27.342 380 72		37.60	-6.18	293.89	0.41	0.50	0.65	0.50	0.55	-5	+2	+3	+7	-19	-8	-17	+3	+8	-10	2	0.91
32323		06 44 52.08	-59 18 13.9	9.53		H	101.216 982 02	-59.303 872 91		1.30	-2.14	7.28	0.72	0.77	0.80	0.79	0.93	-3	-4	+2	+7	+0	-23	+1	+12	-2	-5	0	0.46
32324		06 44 52.52	+19 31 39.5	8.12		H	101.218 853 55	+19.527 631 57		3.61	-4.73	1.98	1.82	1.13	1.79	2.11	1.54	+19	-8	-18	-41	-19	+14	-22	-66	+18	+15	2	
32325		06 44 53.37	+23 13 40.2	8.85		H	101.222 362 82	+23.227 843 36		1.73	-23.86	1.91	1.61	0.70	1.77	1.79	1.11	+16	+56	-2	+2	-5	-6	+17	-52	+37	+17	6	0.57
32326		06 44 53.45	-66 51 56.4	8.71	1	H	101.222 705 66	-66.865 662 50		16.46	-38.05	-48.55	0.69	0.66	0.72	0.49	0.70	-2	+27	+17	-68	-1	-19	+0	-19	-5	-10	2	-0.22
32327		06 44 53.81	+00 37 12.6	9.24		G	101.224 228 42	+00.620 163 64		1.82	1.57	0.72	1.30	0.92	1.33	1.42	1.17	-53	+19	+1	-56	+35	-7	+34	-69	-6	-43	0	-0.11
32328		06 44 54.55	-48 42 28.7	9.18		H	101.227 271 29	-48.707 964 14		6.84	8.98	64.34	0.76	0.72	0.81	0.85	1.01	-9	+11	-1	-17	-1	+9	-2	-14	-7	-9	1	-1.03
32329		06 44 54.86	+20 51 38.2	7.20		H	101.228 599 62	+20.860 622 25		21.77	93.72	12.86	1.02	0.45	1.07	0.98	0.61	+6	+62	+2	-2	+0	-15	+16	-45	+7	+29	0	-0.20
32330		06 44 55.61	-50 34 20.1	9.09		G	101.231 692 88	-50.572 243 64		1.94	-7.26	10.08	0.71	0.66	0.75	0.84	0.85	-8	+22	-13	+9	-8	+4	-9	+3	-5	-9	3	2.03
32331		06 44 55.62	+22 02 01.4	8.63	2	H	101.231 768 59	+22.033 728 98		2.44	0.54	-1.99	1.10	0.52	1.18	1.12	0.69	+3	+51	-1	+8	-2	-10	+14	-43	+15	+17	0	-0.17
32332		06 44 55.67	-70 26 01.6	6.10		H	101.231 974 46	-70.433 772 84		5.18	-15.57	5.94	0.50	0.45	0.49	0.53	0.51	-6	+14	-3	-48	+14	+2	+16	-27	+13	-4	0	0.59
32333		06 44 56.36	+22 50 14.6	6.76		H	101.234 828 69	+22.837 382 14		8.42	-15.80	-0.18	1.04	0.45	1.04	1.18	0.64	+13	+43	+4	+29	+9	-7	+25	-36	+18	+24	0	-0.29
32334		06 44 57.88	-35 07 18.1	8.04		G	101.241 164 23	-35.121 702 79		3.55	-2.34	10.74	0.52	0.58	0.69	0.61	0.75	+2	+10	-14	+6	-7	+1	-10	+23	-5	+11	0	1.35
32335		06 44 59.04	-27 15 59.2	8.90	1	H	101.246 015 57	-27.266 453 67		2.07	-3.67	4.17	0.68	0.88	1.15	0.83	1.02	-16	-1	+0	+7	-14	-14	-14	+4	+5	-18	1	0.35
*32336		06 44 59.18	-67 49 33.6	8.57		H	101.246 577 18	-67.825 994 92		1.44	-12.38	2.71	0.69	0.67	0.72	0.68	0.77	-2	+29	+6	-20	-4	-6	+0	-7	+7	-18	0	-1.00
*32337	H	06 44 59.14	+10 45 09.6	8.00		H	101.246 430 63	+10.752 673 67	*	3.05	2.71	-12.71	1.64	1.05	1.16	1.40	1.05	-32	+22	-1	-24	+14	+2	+17	-67	+5	-25	2	-0.67
32338		06 44 59.76	-05 56 33.6	8.06		H	101.249 010 03	-05.942 678 53		2.13	-0.91	-0.40	0.81	0.59	1.02	1.05	0.85	+3	+38	-12	-15	-24	-11	-14	-32	+13	+33	0	1.04
32339	H	06 45 02.56	-30 35 11.1	6.53		G	101.260 675 92	-30.586 414 96	A	3.51	-7.96	4.44	0.55	0.64	0.73	0.57	0.73	+0	+1	-3	+16	-5	+8	-7	+4	+1	-7	7	-0.33
32340		06 45 03.08	+09 34 18.9	10.64		H																							
32341		06 45 03.65	+45 58 35.3	8.35		H	101.265 223 36	+45.976 477 62		2.02	3.59	-4.57	0.94	0.71	1.16	1.31	0.98	-2	+24	-18	+5	-16	+15	-8	-30	+17	+7	0	1.06
*32342		06 45 05.57	-46 56 09.1	7.78		H	101.273 190 10	-46.935 862 90		3.72	-6.28	10.99	0.57	0.60	0.66	0.74	0.77	-4	-3	-6	-10	+7	+13	+4	-34	+25	-14	0	1.07
*32343		06 45 05.54	-52 21 44.7	7.84		H	101.273 062 89	-52.362 430 36		5.02	-6.03	39.23	0.64	0.55	0.64	0.71	0.63	+7	+6	-7	-9	-10	+5	-10	-19	+21	-9	0	-1.78
32344		06 45 06.32	-11 24 47.8	8.92		H	101.276 321 13	-11.413 269 09		9.32	17.23	-17.13	1.12	0.92	1.26	1.58	1.18	-35	+29	-13	-22	+11	-4	+16	-27	+18	-47	5	0.72
32345		06 45 06.70	-01 01 15.5	7.45		H	101.277 919 88	-01.020 985 94		2.20	-1.14	-1.91	0.77	0.50	0.98	0.87	0.68	-21	+45	-34	-27	-11	-9	-5	-39	+5	+36	0	-1.99
32346		06 45 06.79	+13 45 10.8	8.56		H	101.278 307 25	+13.752 999 45		3.77	-3.68	-8.37	0.94	0.59	1.10	1.06	0.77	-17	+32	-7	-6	+18	-37	+15	-47	-1	-18	0	-0.80
32347		06 45 07.20	-13 12 28.1	9.31		H	101.279 995 54	-13.207 795 33		6.29	-27.11	-3.21	1.20	0.90	1.38	1.44	1.08	+3	+24	-7	-3	-27	-9	-20	-17	+25	-10	0	-0.41
32348		06 45 07.23	-31 39 15.6	7.61		H	101.280 136 83	-31.654 326 31		1.67	-7.30	5.13	0.48	0.52	0.66	0.57	0.57	-3	+10	+5	+22	-20	+2	-18	+18	+1	-7	0	-1.22
32349		06 45 09.25	-16 42 47.3	-1.44	2	G	101.288 541 05	-16.713 143 06	+	379.21	-546.01	-1223.08	1.21	1.04	1.58	1.33	1.24	+5	+22	-30	-50	+13	-31	+12	-54	+32	-5	0	
32350		06 45 09.73	+39 22 06.5	6.93		H	101.290 561 19	+39.368 467 56		18.38	-13.15	-135.45	0.79	0.50	0.86	0.99	0.68	+2	+32	+9	-6	-2	-8	-5	-35	-3	-8	0	-0.69
32351		06 45 09.93	-82 53 45.1	7.60		G	101.291 373 91	-82.895 866 43		12.43	-32.71	9.80	0.55	0.56	0.60	0.62	0.63	-6	+20	+7	+14	+5	+13	+0	-24	+7	-22	1	1.45
32352		06 45 10.01	-18 16 05.7	6.90		H	101.291 721 67	-18.268 244 82		7.65	0.27	-27.46	0.48	0.51	0.87	0.60	0.68	+0	+11	-7	+29	+38	-6	+37	-14	+10	-8	0	-0.98
32353	H	06 45 10.02	+30 49 33.5	7.27		H	101.291 754 57	+30.825 973 01	A	14.41	-10.34	-15.49	1.46	0.81	1.37	1.66	0.98	+15	+25	-27	+0	-19	+18	-8	-48	+40	+24	3	1.23
32354		06 45 10.49	-47 13 21.8	7.22	2	H	101.293 703 55	-47.222 733 54		2.69	-1.53	6.55	0.54	0.57	0.62	0.77	0.76	+3	-4	+1	-9	+7	+23	+2	-34	+17	-7	0	-0.38
32355		06 45 11.21	-00 02 35.4	7.02		G	101.296 706 32	-00.043 173 63		1.90	0.40	-1.71	0.71	0.48	0.87	0.80	0.61	-32	+25	-29	-41	-1	+8	+6	-51	+19	+2	0	-0.70
32356		06 45 13.35	-46 51 12.1	7.49		G	101.305 614 09	-46.853 371 41		3.29	-6.27	11.64	0.55	0.57	0.62	0.72	0.73	-5	+5	-6	-10	+12	+12	+9	-37	+16	-18	0	1.61
32357		06 45 13.85	+05 12 47.7	7.53		H	101.307 707 36	+05.213 263 81		4.58	-11.77	3.57	0.99	0.62	1.03	0.98	0.78	-17	+24	+2	-43	+4	-14	+10	-66	-24	+14	0	-1.11
32358		06 45 14.46	+53 05 10.0	7.04	1	H	101.310 233 15	+53.086 115 14		2.58	-14.49	-21.48	0.67	0.48	0.83	0.74	0.53	-12	+31	-3	+24	+34	+2	+30	-14	+12	-26	0	0.13
32359		06 45 14.58	+13 20 09.4	7.28	1	H	101.310 748 60	+13.335 933 34		1.44	-0.51	-0.57	0.81	0.45	0.89	0.85	0.60	-6	+49	-8	-4	+1	-21	+9	-48	+6	+10	0	-3.02
32360		06 45 14.66	-25 38 24.5	8.40		H	101.311 082 82	-25.640 144 89		0.03	-3.31	3.43	0.52	0.66	0.95	0.61	0.79	+8	+11	+2	+13	-19	+8	-20	-11	-1	+9	2	-1.44
32361		06 45 17.17	+46 51 52.8	8.81		H	101.321 540 91	+46.864 654 84		6.97	-24.86	0.83	1.03	0.71	1.09	1.46	0.90	+12	+25	+2	-7	-13	+5	-12	-31	-4	+16	0	-0.20
32362		06 45 17.43	+12 53 45.8	3.35		H	101.322 639 80	+12.896 055 13		57.02	-115.16	-190.91	0.74	0.43	0.83	0.67	0.50	-10	+54	-10	-15	+7	-31	+4	-43	-7	+8	0	-1.42
32363		06 45 17.51	+82 25 01.4	8.63		H	101.322 968 28	+82.417 042 07		1.71	0.59	-6.14	0.70	0.74	0.92	0.74	0.90	+20	+31	+12	-15	+0	+15	-2	+1	+5	+30	1	-0.63
32364		06 45 21.89	+35 32 06.3	8.82		H	101.341 217 10	+35.535 086 55		18.76	84.57	-210.32	1.59	1.07	1.44	1.88	1.28	+39	+13	-26	-60	-40	+26	-32	-83	+28	+32	0	-0.64
32365		06 45 22.68	-34 37 41.2	9.06		H	101.344 505 61	-34.628 118 31		3.47	-3.77	3.47	0.67	0.83	0.96	0.78	1.16	-4	+10	-23	+12	-4	+17	-2	+21	+4	+10	1	2.39
32366		06 45 23.08	-31 47 34.5	5.92		G	101.346 167 95	-31.792 921 74		41.20	-204.91	-304.13	0.40	0.43	0.56	0.48	0.48	+9	+18	+9	+14	-11	+7	-11	+14	-1	+7	0	0.06
32367		06 45 23.28	+23 38 46.4	7.16		H	101.346 991 57	+23.646 229 46		4.02	24.18	0.04	1.03	0.50	1.05	1.21	0.70	+22	+37	+1	+19	-3	-4	+11	-36	+19	+33	2	-0.22
32368		06 45 23.38	-23 27 45.0	6.04		H	101.347 405 23	-23.462 488 65		6.63	-19.70	23.38	0.34	0.42	0.64	0.37	0.45	+4	+10	-7	+41	-32	+14	-22	-10	+18	+22	1	-0.61
32369		06 45 23.51	+24 40 20.9	6.95	1	H	101.347 942 30	+24.672 478 08		0.62	-0.01	-3.17	1.00	0.54	1.02	1.06	0.64	+21	+41	+10	+9	-10	+9	+1	-40	+9	+35	0	1.95
32370		06 45 24.28	-09 58 48.9	8.45		G	101.351 154 38	-09.980 251 56		3.90	-0.41	-4.41	0.94	0.72	1.02	1.09	0.87	-20	+16	+4	-38	+10	+9	+10	-48	+0	-14	2	1.05
32371		06 45 24.50	-26 09 21.0	9.12		H	101.352 066 94	-26.155 823 35		3.83	-14.42	-3.68	0.71	0.86	1.17	0.87	1.08	-12	-10	+3	+5	-11	+8	-10	-2	-3	+0	0	0.49
32372		06 45 24.58	-28 26 16.0	9.53		H	101.352 419 43	-28.437 789 13		4.16	10.58	3.69	0.96	1.13	1.56	1.20	1.38	-4	+0	+11	+10	-10	+2	-8	-8	-12	+6	0	1.49
32373		06 45 25.15	+47 04 01.5	8.28	1	G	101.354 797 39	+47.067 091 47		3.32	8.94	-11.53	1.77	1.28	1.96	2.32	1.44	-4	+21	-13	-12	-7	+4	-3	-41	+0	+9	0	
32374		06 45 25.28	-65 02 39.2	7.18	2	H	101.355 345 60	-65.044 225 38		2.83	-8.38	-5.80	0.50	0.50	0.52	0.56	0.56	+2	+3	+9	+5	-13	+6	-14	+2	+12	-4	0	0.55
32375		06 45 26.05	-52 12 03.4	6.56	1	G	101.358 555 46	-52.200 931 17		1.49	2.19	2.71	0.50	0.46	0.52	0.58	0.52	+8	+13	+8	-1	-14	-6	-13	-8	+13	-9	1	-0.64
32376		06 45 26.51	+09 45 56.9	6.65		H	101.360 465 17	+09.765 812 72		11.03	41.55	-32.30	0.81	0.50	0.89	0.98	0.72	+7	+27	+18	-9	-12	-23	-3	-65	-28	+27	0	1.44
*32377		06 45 27.04	-17 50 31.1	9.56		H	101.362 669 21	-17.841 965 95		2.65	4.15	5.50	0.92	1.09	1.76	1.20	1.40	-15	+12	-31	+13	+40	-33	+32	-38	+30	-30	0	0.67
*32378	H	06 45 27.02	+29 21 59.8	8.22		G	101.362 577 08	+29.366 618 61	A	11.07	-32.30	-74.61	3.70	2.58	1.74	3.29	2.70	-18	+22	-6	-28	+47	-1	+55	-57	+20	-58	0	-0.75
32379		06 45 27.67	+38 22 57.6	8.02	1	H	101.365 305 89	+38.382 664 01		7.87	-13.76	-32.74	0.93	0.61	1.09	1.20	0.75	-9	+27	-28	-7	-22	+24	-3	-52	+22	+23	0	-0.52
32380		06 45 27.86	-07 08 10.2	8.44		H	101.366 081 97	-07.136 180 22		1.63	-4.89	8.19	0.92	0.69	1.13	0.97	0.81	-9	+41	-34	-33	-23	-11	-18	-37	+23	+7	0	-1.97
32381		06 45 28.78	-36 50 57.9	7.98		H	101.369 899 88	-36.849 420 81		6.04	-4.44	-7.71	0.56	0.62	0.73	0.63	0.76	+15	+3	+5	+16	+5	+21	+4	+27	+2	+15	1	0.67
32382		06 45 28.86	+44 14 01.8	7.47		G	101.370 257 83	+44.233 847 16		22.09	143.37	-189.41	0.74	0.56	0.86	0.97	0.67	-4	+3	-4	+2	-24	-2	-22	-21	+13	-9	0	-0.43
32383		06 45 30.50	+15 28 48.0	8.43		H	101.377 068 44	+15.479 993 86		2.75	0.47	-10.37	1.10	0.55	1.18	1.10	0.70	-11	+52	-25	+7	-1	-16	+14	-46	+12	+23	0	1.61
32384		06 45 30.94	-01 25 47.9	7.82		H	101.378 909 23	-01.429 984 20		1.99	-1.47	-2.74	0.74	0.53	0.87	0.79	0.61	+1	+18	-21	-31	-17	-1	-16	-44	+7	+14	0	0.97
32385	H	06 45 31.19	-30 56 56.4	5.62	2	H	101.379 961 61	-30.948 990 76	A	2.06	-3.11	4.20	0.70	0.78	0.90	0.78	0.87	+7	-23	-26	-14	-9	+16	-14	-37	+25	-8	17	0.54
32386		06 45 32.30	-14 57 26.5	8.95		H	101.384 574 03	-14.957 373 37		4.04	-6.43	3.06	1.09	1.01	1.41	1.65	1.38	-16	+28	-34	-7	-10	-1	-1	-24	+31	-31	0	1.43
32387		06 45 33.10	+04 35 38.9	7.04		H	101.387 903 70	+04.594 130 93		2.66	-12.57	-4.64	0.78	0.51	0.86	0.86	0.64	+7	+22	+2	+1	-10	-5	+1	-46	-13	+35	0	0.10
32388	H	06 45 33.95	-31 30 03.0	8.93		H	101.391 452 39	-31.500 840 45	A	6.22	1.23	-8.39	1.06	1.10	1.29	1.19	1.23	-7	+4	+11	+31	-2	-10	-6	+1	+12	-21	4	0.46
32389		06 45 35.10	+27 40 23.7	6.70		H	101.396 247 38	+27.673 248 53		3.59	-5.28	-0.89	0.88	0.52	0.96	1.17	0.80	-6	+19	-9	+6	+16	+5	+19	-50	+16	-6	0	0.08
32390		06 45 35.52	-20 40 51.4	8.39		G	101.397 982 20	-20.680 934 26		0.38	-5.36	-0.38	0.65	0.74	1.08	0.90	0.94	+2	+12	-10	+19	+12	-2	+11	-31	+6	+3	1	-0.19
32391		06 45 37.69	-24 11 01.2	8.82		H	101.407 059 35	-24.183 665 04		3.98	6.89	10.83	0.55	0.71	1.05	0.66	0.74	+7	+7	+0	+19	-28	+13	-23	-12	+18	+11	0	-0.64
32392		06 45 42.52	-66 20 33.4	8.87		H	101.427 174 48	-66.342 603 95		2.13	-3.29	16.33	0.76	0.72	0.79	0.61	0.80	-1	+16	+13	-56	-4	-9	+1	-3	-5	-14	0	0.08
32393		06 45 43.01	-20 51 09.6	7.44		H	101.429 227 74	-20.852 664 23		2.18	-4.73	0.77	0.59	0.65	0.97	0.79	0.82	-3	+15	-7	+9	+14	+3	+13	-33	+10	-6	0	0.09
32394		06 45 46.40	+37 46 41.4	8.02		H	101.443 336 50	+37.778 180 12		0.61	3.21	-11.71	0.94	0.60	1.09	1.16	0.73	-9	+26	-12	-9	-15	+8	-3	-53	+5	+15	0	-0.04
32395		06 45 48.46	-45 59 51.9	9.00		H	101.451 929 91	-45.997 737 14		11.89	77.31	-1.18	0.66	0.73	0.76	0.86	0.98	-11	+0	+0	-4	+20	-6	+14	-29	+7	-37	0	-0.87
32396		06 45 49.74	-70 31 21.6	7.60		H	101.457 231 97	-70.522 652 79		5.91	18.35	-10.66	0.64	0.56	0.62	0.66	0.65	+0	+19	-2	-51	+9	+1	+13	-27	+19	+3	0	0.27
32397		06 45 49.98	+02 29 57.4	7.22	2	H	101.458 241 96	+02.499 282 46		1.11	-1.77	-2.51	0.82	0.54	0.88	0.86	0.69	+12	+30	-15	-43	-22	-20	-24	-56	-6	+33	0	-1.09
32398		06 45 50.06	-00 42 59.2	6.78		H	101.458 571 45	-00.716 448 83		4.96	-5.90	-4.20	0.73	0.46	0.88	0.80	0.63	-14	+40	-19	-32	-7	-11	-2	-43	+2	+32	0	0.50
32399		06 45 50.91	+62 38 58.5	7.56		H	101.462 145 42	+62.649 572 38		4.23	-0.87	3.41	0.62	0.59	0.94	0.92	0.81	+30	+13	-5	+8	-12	+8	-4	+5	+34	+19	0	-0.84
32400	H	06 45 51.85	+03 24 02.5	8.86		H	101.466 025 74	+03.400 696 07	A	-0.74	-2.70	1.82	3.70	2.44	3.29	4.40	3.30	-27	+19	+7	-25	-5	-33	-8	-51	-8	-5	2	0.79

Hans Schrijver

65195 | 00437

One of more than two thousand pages of the Hipparcos catalogue (the page contains the bright star Sirius, Hipparcos catalogue number 32349)

Chapter 10

The Finishing Touches

For my part I know nothing with any certainty, but the sight of the stars makes me dream.

Vincent van Gogh (1853–1890)

ALTHOUGH THE HIPPARCOS and Tycho catalogues had been completed in 1997, several other important pieces of this grand endeavour were nevertheless still marching on.

Erik Høg's enormous work in assembling the Tycho catalogue took a new turn. Immediately after its publication, he and his small team in Copenhagen started a reprocessing of the satellite data to dig out the signals of some even fainter stars that had not been extracted in the first pass. This renewed effort took the catalogue from a million stars to more than two and a half million.

A related, but even bolder task was taking shape: Høg, amongst others, had recognised that the star positions from the Astrographic Catalogue, the century old survey of ground-based positions, could be combined with the space measurements, to give accurate motions for all stars in his reprocessed and enlarged Tycho catalogue. Between 1987 and 1994, astronomers at the Sternberg Institute in Moscow, led by Andrei Kuzmin, had created a computer-readable version of the Astrographic Catalogue by scanning the two hundred and fifty four printed volumes left by that project as its gift to science in the first half of the nineteenth century. The task was a huge one, the main problem being to check the integrity of the mountain of numbers that had been scanned, digitised and stored electronically.

M. Perryman, *The Making of History's Greatest Star Map*, Astronomers' Universe
DOI 10.1007/978-3-642-11602-5_11, © Springer-Verlag Berlin Heidelberg 2010

Erik Høg joined forces with collaborators from the US Naval Observatory in Washington, led by Sean Urban. Under Naval Meteorology and Oceanography Command, the observatory is the official source of time for the US Department of Defense, and for the entire United States. Their contributed expertise was in the handling and calibration of other large deep sky surveys. Applying their proficiency to these two catalogues—the old positions from the Astrographic Catalogue, and the deep pass of the Tycho stream from the Hipparcos satellite a hundred years later—Høg and collaborators built the Tycho 2 catalogue: two and half million stars, each with proper motions, constructed over nearly one hundred years of elapsed time. Although the old observations were not as accurate as the new, their added strength was the grand lever arm of time which gave added power to the detection of the star motions. The outcome was a very large catalogue with proper motions roughly comparable in accuracy to those from the Hipparcos catalogue itself. While of limited interest for parallax distances, it has still proven to be an immensely valuable source of accurate positions and motions for an even larger assembly of stars than the Hipparcos catalogue itself.

Meanwhile, Floor van Leeuwen in Cambridge continued to unravel a deeper understanding of the satellite's complex spinning motion in space, another marathon effort which led to an improved Hipparcos catalogue appearing in the world's astronomy data centres in 2007.

THE MILLENNIUM STAR ATLAS was our foray into the field of celestial cartography, or uranography, the branch of astronomy concerned with pictorially mapping the stars. The idea was to transform the printed and computer catalogue listings into true celestial charts.

Bayer's Uranometria: Orion

With the grand atlases of history to guide us, we hoped to create something of aesthetic appeal too. Early star atlases, based on naked-eye measurements, were actually primarily works of science, providing accurate star charts on which to plot the changing positions of planets, comets, and the Moon. Of course they were frequently remarkable works of art. Amongst the most notable have been the generally-accepted 'grand' celestial atlases of Johan Bayer's *Uranometria* (Augsburg, 1603), Julius Schiller's *Coelum Stellatum Christianum* (Augsburg, 1627), Johannes Hevelius's *Firmamentum Sobiescianum sive Uranographia* (Gdansk, 1690), John Flamsteed's *Atlas Coelestis* (London, 1729), and Johann Bode's *Uranographia* (Berlin, 1801).

The Millennium Star Atlas was a pioneering collaboration between the Hipparcos project, and a team at Sky Publishing led by Roger Sinnott. Rick Fienberg, Leif Robinson, and Roger Sinnott, from *Sky & Telescope*, whose names will be familiar to readers of this leading astronomy magazine, had approached me before the catalogues were finalised to sound me out about embarking on such an undertaking. Roger had thought the project through in some detail, and knew what was possible and what was needed. It was clearly a highly attractive proposition: the quality of the numerical catalogue would shine irrespective, but a visual representation would augment the dry, telephone-directory style austerity of a printed catalogue alone. The professional astronomers would use the on-line data, but the large and discerning amateur community would value the atlas. We threw ideas back and forth about what it could and should contain, sketched out some concepts, quickly reached agreement, and got down to work.

Roger Sinnott (2007)

From Europe, we sent over preliminary catalogue data to test out the interfaces, and once the final catalogue was completed, shipped it in its entirety to the Sky Publishing team. Proofs of the charts were shipped back from the USA for us to verify, and the final star maps, glittering like the jewels they contained, were delivered by installments into the major publishing pipeline that the publications department in ESA was undertaking. Roger Sinnott later joined us at the forthcoming Venice Symposium to celebrate the catalogue's publication, another welcome guest, presenting his highly valued role in these final steps.

With the catalogue publication behind us, I made a visit to Boston in January 1998, only then to meet for the first time the team that we'd worked so intensively and successfully with to create the Atlas. It was a great pleasure to work with Roger, whose knowledge of the night sky was impressive—Alnitak and Alnilam, Castor and Pollux, Deneb and Denebola, all were his friends.

Millennium Star Atlas team (1997)

It was a splendid collaboration, each side pooling their unique expertise, each respecting the other's.

The Millennium Star Atlas was a grand undertaking, and well received by the amateur astronomy community, although its price was high. It extended earlier attempts at completeness and uniformity to a limit of around 10–11 magnitudes. Just over fifteen hundred charts include a million stars from the Hipparcos and Tycho catalogues, three times as many as in any previous all-sky atlas; more than eight thousand galaxies; numerous bright and dark nebulae; open and globular clusters; and two hundred and fifty of the brightest quasars.

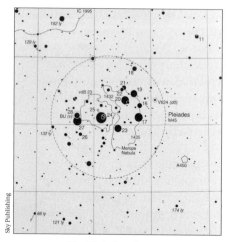

*The Pleiades cluster
in the Millennium Star Atlas*

Distance labels are given for stars within two hundred light-years of the Sun. Proper motion arrows are given for stars with motions exceeding two tenths of a second of arc a year. Variable stars are characterised by amplitude and variability type. Thousands of already known and newly-discovered Hipparcos double stars have tick marks showing their separation and orientation on the sky.

William Liller, former chair of the astronomy department at Harvard, eulogised: "The Atlas is, simply put, magnificent. The enormous number of highly accurate positions, magnitude measurements, and greatly improved parallaxes and proper motions have made it possible to produce an atlas like never produced before." History has yet to judge the Millennium Star Atlas in context, but we may hope that it will be viewed favourably.

Later, Roger Sinnott and I experimented with a version of the atlas in which the star colours were faithfully reproduced on a black background, and thus more closely representing the true grandeur of the night sky. The prototype was stunning, but publication proved prohibitively pricey.

The Millennium Star Atlas was the first to include the Hipparcos and Tycho catalogues, although all other major popular sky atlases since 1997 have also incorporated their findings. Star chart guru Wil Tirion in The Netherlands accordingly led the creation of the second edition of Sky Atlas 2000 in 1998, the third edition of the Cambridge Star Atlas, and the second edition of Uranometria 2000, amongst others. Planetarium software around the world, amateur and professional, swiftly incorporated the results into their data bases, with 'Sky in Google Earth' following in August 2007.

IF THE MILLENNIUM STAR ATLAS was an important way of portraying the results, I gambled on one with greater visual impact after being invited to give the Royal Astronomical Society's George Darwin Lecture[41] in 1998 at their grand premises in London's Piccadilly. I searched for a way to capture the impact of what we had measured.

I settled on a stereoscopic projection of the night sky in a darkened auditorium, my own planetarium. I set about trying to capture, in a lecture theatre, the three-dimensional distribution of stars in space. The solution was to project two images of the sky, as if seen from the two extremities of the Earth's orbit. Using polarised light, in the way that has been used intermittently for stereo movies over many decades, the two eyes see the view of space from slightly different directions. The brain fuses the two images into a stereoscopic perception, in exactly the same way as our eyes, slightly separated, capture a truly three-dimensional view of the world around us.

The technology available at the time of the catalogue publication was somewhat limited, and I settled on using two 35 mm slide projectors, with a remote control that would advance, synchronised together, the pair of slides needed to give the three-dimensional effect.

The entire enterprise needed a special projection screen some meters in size, with a metallic coating to preserve the polarisation. I needed a couple of hundred polarised glasses for the audience. My colleague Karen O'Flaherty helped with the preparations and with the set-up in London where, despite our combined conviction that such a temperamental system of synchronised plastic slide projections was bound to go disastrously wrong and jam, it actually proceeded without a hitch. A hundred and sixty years after the first parallax distance had been presented to the Royal Astronomical Society, its members could sit and gaze at a replica of space as it would appear in its full three-dimensional glory.

Relief and elation at escaping without the brute force slide projection failing dismally, which would have made the lecture memorable but for the wrong reason, turned to a mixture of pleasure and not a little trepidation when, less than a year later, the International Astronomical Union issued an invitation for me to give its plenary Invited Discourse at the forthcoming General Assembly to be held in Manchester in 2000. "We would be delighted, but do not insist," the invitation read, "if you would consider repeating the three-dimensional projection on this occasion." Word had got around. It would be an even bigger gamble to accept, but certainly an opportunity not to be missed.

The two slide projector approach could be discarded immediately. The lecture was to be in the Bridgewater Hall in Manchester, the spectacular auditorium created by architects Renton Howard Wood Levin and the residence, since 1996, of the Hallé Orchestra. I could expect an audience of more than a thousand, and proportionally more pressure for the technology not to fail. The solution was two industrial-strength high-power projectors, each of which required four people to lift, positioned towards the rear of the concert hall, and carefully aligned to give the projected stereoscopic effect. An imposing six-meter wide screen was nevertheless almost lost on stage in the huge grandeur of the auditorium.

Viewing the stars in three dimensions

I was able to exploit the rapid progress of video technology over the space of just two years to include the movements of the stars as well as their distances. Jos de Bruijne, one of the new generation of European experts who had cut his own astrometric teeth with a PhD making use of the Hipparcos catalogue, helped with some stunning stereo visuals. This gave other things that might go wrong, but I made a dry-run to make sure that it didn't. In the event, more than a thousand astronomers sat through my presentation in August 2000. Most were more enthused by the power of astrometry when seeing, almost deity-like, the motions of stars through the three-dimensional grandeur of space.

As a postscript, the technology has moved on and the three-dimensional videos can now be played out on a laptop, although powerful projectors, a dark auditorium, and a big special reflecting screen are still needed to replicate the stunning stereoscopic effects for a large audience. It was perhaps this kind of mental image, and just how this knowledge might help us understand our Galaxy, that drove many of us on to map the stars. Stereoscopic viewing will surely soon descend upon our entertainment technology, television and cinema, and visualising the stars in three-dimensions will probably become a more widespread possibility.

Until then, I can only ask you to imagine staring up at the night sky, and sensing the awe-inspiring impact of the stars not simply as pinpoints of light scattered across the uniform surface of the celestial sphere, but as a profusion of suns extending around us in three dimensions, ranging outwards to infinity, in timeless motion.

IN THE MEANTIME, the final data was sent to the two hundred proposers who had a year of priority access, and who could settle down to the science investigations they had outlined a decade before.

A couple of special cases were given early access. The very first was a request from NASA's interplanetary navigation team at the Jet Propulsion Laboratory in Pasadena to know accurate positions of a few stars in one specific direction in the sky, so that they could optimise the navigation of their Galileo spacecraft, now hurtling outwards on its way to Jupiter. Passing twice through the aster-

Ida and moon Dactyl from Galileo

oid belt, on the first pass they took in a rendezvous flyby with asteroid Gaspra in 1992 from a distance of 1600 kilometers and, on the second, with asteroid Ida and its own moon Dactyl in 1993. We handed over the data with confident anticipation; their spectacular close-up images of these distant solar system objects were soon splashed across the world's press, and NASA duly posted off certificates of achievement for our part in the fly-by success. It's the professional scientist's equivalent of a gold star at primary school, but still rather pleasing to receive.

Richard West, from the European Southern Observatory's headquarters in Munich, made a somewhat similar petition a couple of years later. He was actively orchestrating a multi-national ground-based observing programme of the imminent demise of Comet Shoemaker–Levy. The comet's orbit had been carefully tracked over previous apparitions, and was now predicted to collide with Jupiter in 1994. This head-on cosmic impact would provide the first direct opportunity to observe an extraterrestrial collision of solar system objects, and a battery of telescopes would be

Comet Shoemaker–Levy impacting Jupiter in 1994

pointed in concert to Jupiter to record the event. The comet had already been torn apart by the gravitational forces of a previous near-miss encounter with Jupiter in 1992. These disrupted fragments were still strung out like pearls on an invisible celestial necklace, and they finally collided with Jupiter, one after another, between 16–22 July 1994 at an impact speed of sixty kilometers a second. Our accurate star positions gave an unprecedented reference frame which allowed the expected position of Jupiter and the cosmic trajectory of the comet to be perfectly aligned. Better estimates of their times of impact allowed for closely-synchronised observation by astronomers worldwide.

The spectacular impacts, like large thermonuclear bombs, generated large coverage in the popular media, and the collision provided new information about Jupiter and highlighted its role in reducing space debris in the inner solar system. Sucked up by Jupiter's giant gravitational field like a celestial vacuum cleaner, our gas giant neighbour seems to play a valuable role in reducing potential impacts of rogue asteroids with Earth itself. It may be that, without this colossus guarding us at the portals of the inner solar system, life on Earth might have developed very differently, or perhaps even have been repeatedly wiped out along its precarious path.

AS NEWS OF THE CATALOGUE'S imminent release leaked out, and word of the quality of the contents spread, so Hipparcos found itself once more in the world's press, this time for all the right reasons. At last we had our fifteen minutes of fame. Although journalists are so often knocked for their sensationalism or their economies of fact, I found the reports in the British press—I cannot speak objectively for the others—to be almost invariably accurate and well balanced, from the pain of 1989 to the elation of the mid-1990s. Amongst others Mary Fagan in The Independent, Roger Highfield and Steve Connor in The Daily Telegraph, Tim Radford in The Guardian, Bruce Dorminey in The Financial Times, and Nicholas Booth in the Observer, all produced well-crafted pieces always accurate in their coverage.

BBC television's technology feature *Tomorrow's World* filmed the team in action in the main control room of operations centre at ESOC in December 1989 where Howard Stableford nicely conveyed the actuality and the expectation. Sir Patrick Moore and his *Sky at Night* team visited ESA's technology centre in Noordwijk in February 1992 and filmed in front of the satellite engineering model now hanging in the Space Exposition there. Patrick's enthusiasm and admiration for the work of professional astronomers shone throughout his visit.

Michael Feast (2007)

Michael Feast, an astronomer from South Africa, was one of the first to publish results from the early harvest. He had been studying the first Hipparcos distances of the important class of Cepheid variable stars. Up until that time there was an awkward and unsettling paradox regarding the age of the Universe. From its expansion, its age had been estimated at no more than about eleven billion years, while some of the stars within it had ages, estimated from theories of their luminosity and their evolution, of around fifteen billion years. Clearly something, somewhere was wrong for science to be admonishing us that some of the objects in the Universe were older than the Universe itself. Astronomical distance estimates use a sort of ladder of comparisons which stretch out from stars in the solar neighbourhood to objects billions of light years away. Feast's swift analysis nailed down the first steps on the rickety ladder a little better. This brought stellar ages down to about eleven billion years, and pushed the Universe's up a little, to around twelve billion years. With consistency established for the first time, astronomers could breath more easily that two foundations of their science—cosmology and stellar evolution—were not, after all, incompatible.

It's not often a scientist will get to resolve such a high-profile paradox, to wipe a couple of billion years off the face of a cosmic timepiece, or to add a billion years, give or take, to the age of the Universe. He was rushed by taxi from a meeting of the Royal Astronomical Society in London to an interview for BBC radio's *Science in Action.* In an unexpected juxtaposition of writer and newspaper for still a rather arcane physics experiment, Andrew Derrington, professor of psychology at Nottingham, picked up the story and wrote a half-page feature in the Financial Times, lucidly explaining what was going on[42]. He likened it to the problem he had faced when, at the age of five, he had been unable to figure out the age of adults from their sizes relative to his brothers and sisters. "For many years I believed my father to be 104 years old," he wrote. "Even after I discovered that *his* father was less than 70, it took me a long time to resolve the paradox."

THE MAIN HIPPARCOS CATALOGUE of a hundred and twenty thousand stars along with the first of the Tycho catalogues of more than a million were finalised and published early in 1997. We decided on an international symposium to present the results.

Several of the project's leading scientists expressed a keen interest to host the symposium in their home countries, to bathe in the reflected glory of the measured starlight. Proposals were formulated and presented. A straw poll sealed the fate of losing bids from Paris, London, and Noordwijk in The Netherlands. Venice was, perhaps not too surprisingly, selected.

Isola di San Giorgio

The unveiling ceremony was held in May 1997 at an international symposium which took place on Isola di San Giorgio, a small island in the lagoon in Venice. The conference centre was within the former Monastery of San Giorgio, part of the cultural Cini Foundation. Venice, of course, turned out to be an incomparable setting. Two hundred and fifty scientists from around the world came to see the catalogue presented, and to hear of the first scientific investigations carried out. Local arrangements were masterminded by Italian science team member Pier Luigi Bernacca from Asiago, who organised a suitably lavish event, certainly by scientific conference standards. Speeches included a welcome from the Mayor of Venice, and an address by Roger–Maurice Bonnet, still ESA's Director of Science, at last able to celebrate the final victory. International Astronomical Union president Lo Woltjer attended, and we welcomed American astrometrists who had crossed the Atlantic to join the acclaim.

Pier Luigi Bernacca (1997)

On the podium for the opening ceremony were Lindegren, Høg, Turon and Kovalevsky as the science consortium leaders, and Roger–Maurice Bonnet, Hamid Hassan and myself from ESA. Industry leaders Michel Bouffard of Matra Marconi Space and Bruno Strim of Alenia Spazio were there to share this formal end to the project, recipients of respect well merited in having turned scientific dreams into technological reality. Years later, Bouffard was to recall the symposium as a highlight of his own involvement in the project which had stood for seven years of his professional life, and which had propeled his own career onward to manage ESA's SOHO solar observatory, and later as the head of EADS Astrium's science and Earth observation directorate.

Everywhere, the excitement was palpable. Coffee breaks were periods of intense discussions between elated scientists, wandering the blossom-filled gardens of the former Benedictine monastery in the warm spring sunshine. In continuing testament to the restlessness of scientific enquiry, the final session of the week was devoted to the future of space astrometry, of star positions after Hipparcos, where vastly more ambitious plans were slowly but surely taking shape.

A gala dinner was held in the fifteenth century Palazzo Pisani della Moretta on the Grand Canal, already a century old when Tycho Brahe was born. Conference delegates arrived by gondola, and departed some hours later, satisfied with the meal, and with the presentations which promised so much more over the years to come.

ESA

Symposium participants, Venice (1997)

Adriaan Blaauw, chair of the observing programme committee fifteen years earlier, was one of the guests of honour. Now a sprightly 83, he was as excited as any of us to see the outcome, keen to understand the implications for young star forming regions, a research topic he had pioneered more than half a century before. In deference to his seniority, he had been assigned his own comely escort from the Cini Foundation's conference staff, but somehow always managed to be sided by two.

We stood on the quayside together one morning, waiting for the ferry to take us over to the cloistered isolation of San Giorgio, and I asked if he had been to Venice before. "Yes," he said, "I cycled here once, from Holland, sixty years ago." As he sprang aboard the boat, striking up conversation with others from his committee who had also come to see the fruits of their own labours a decade before, I would not have been all that surprised if he'd announced that he was cycling back after the symposium ended.

DISTRIBUTED TO THE world's scientific libraries in 1997, the results are freely accessible through astronomical data centres, and repeatedly tapped into for countless investigations. The star positions are now routinely used to direct ground-based telescopes to their chosen targets, point and navigate space missions, and drive public planetaria. As a catalogue of star positions as they were arranged in the sky around 1990, it maps a configuration which will never be seen again. As such, astrometry bestows a rare treasure in science; it provides a unique snapshot of the Universe at one moment, preserved for all time. It may diminish in relevance as future star catalogues from space build upon it, but its historical value will persist indefinitely.

"Altogether thirty years elapsed before our work was completed," said Jean Kovalevsky at an ESA award ceremony in 1999. "For individuals involved from the beginning, it was an extraordinary commitment within a human lifetime. Yet thirty years was a short time in the history of science to achieve a revolution that has affected every branch of astronomy."

Chapter 10

I FINISH THIS CHAPTER with a retrospective on the scientific and technological progress that accompanied the advance of Hipparcos during the 1980s
and 1990s. Although I sat in the middle and conducted its progress, it still
seems curious, even fortuitous, that it all came together. Let me explain.

Satellites, or at least many of the components or technologies used
within them, are destined to be almost obsolete by the time they are
launched. Designs must typically be based on proven, and even space-
qualified technology, which means that a satellite designed in the early
1980s will no longer be carrying state-of-the-art instrumentation, at least
by ground-based standards, by the time it is placed in orbit ten years later.
Satellite design is always a delicate compromise between choosing the most
advanced technology available, or choosing that most likely to work once
in orbit. Too long a gestation period, and the construction lags behind the
technological leaps going on around it. Of course, the sum total of what the
satellite is put into orbit to do should still be state-of-the-art.

For Hipparcos, the detector technology employed—photomultipliers
and image dissector tubes—was out-dated by the time of its launch in 1989.
CCD detectors were already far more advanced in terms of their efficiency,
although their sensitivity to the effects of radiation would probably still have
made them a poor choice even a decade later. The on-board computer was
another striking example of inexorably advancing technological progress:
the total memory size was a mere 64 kilobytes, barely enough to store one
tenth of a single digital camera image today. It had to calculate the attitude,
point the detectors, instruct the gas jets to fire at the correct times, and reor-
ganise the data into a form suitable for transmission to ground. Reprogramming it by commands from ground, to take account of the non-intended
orbit and, later, for the failing gyroscopes, was a touch-and-go exercise in
which every bit and byte of available memory had to count.

When the Hipparcos mission was being designed in the late 1970s, computer power for the data processing on ground was far behind what it was in
the 1990s. I think it is correct to say that the satellite data could not have
been processed on the ground if the technology available after launch had
not progressed by unimagined leaps and bounds in the intervening fifteen
years. The early concepts somehow rested on the implicit assumptions of a
continued evolution of processing and data storage, the doubling of power
every eighteen months or so known in the computer world as Moore's Law.

If they had not, we would have been in a tricky situation. As it was, the
computers on which the analysis was carried out were physically very substantial machines, one of the team's occupying a cavernous vault in the CNES
facility in Toulouse. They were fed their diet by tape robots, the information
stored on thousands of nine-track magnetic tapes.

The numerical processing would have been greatly simplified if it had been carried out even five years later: systems were by then far more compact, and storage considerably more efficient. The sudden emergence of the data highways would have vastly facilitated the data transfer between institutes, and provided the possibilities for improved iterations of the global solutions. But perhaps with more process-

Hipparcos computers, CNES (c. 1990)

ing power on hand, the plans would just have been laid down with less rigour, with less care and attention given to the requirements of each step.

In the early days of the project, no deep thought was given to the problem of how the final results would eventually be made available. It was a problem to be confronted two decades hence. There was, again, an implicit assumption that technologies would have advanced enough for a solution to be found. In the early 1980s, as the satellite was being constructed, there was no world-wide web. CDs or DVDs did not exist, and had clearly not been conceived of as possible media for the published catalogues. Yet the necessary technology somehow kept pace with the project's needs. Or perhaps it was simply a case of adapting to the most advanced solutions available, whatever they would turn out to be. Earlier catalogues had, after all, been circulated on magnetic tape and, before that, on paper. Neither would have been elegant solutions, but both would have worked.

It seems astonishing now, almost comical, to reflect that the observing proposals came in, in the early 1980s, on half-inch nine-track magnetic tapes, as well as on punched cards. The two hundred proposals together weighed in at more than a hundred kilogrammes. In less than two decades, these media have disappeared, long since banished to museums of science and technology.

Almost the entire project was conducted in the days preceding electronic mail. Essentially until the time of launch, communication between the highly distributed European teams working on the satellite design and construction, and the scientific teams working on the starting catalogue and the data analysis, communicated by postal mail. In more urgent cases, confirming a scheduled meeting, or distributing critical information, telexes were used somewhat routinely, even until just after launch.

Face-to-face meetings were held with regularity; the ESA project team would fly en masse to the premises of the prime contractor to review progress perhaps once a month. Scientific consortia meetings were held three or four times a year. The science team gathered on a similar frequency. Participation by project team, industry, or ESOC representation was the principal medium for orchestrating the advance of the project as a whole. Telephone calls were relatively infrequent, especially in the early years, due to a combination of expense and difficulty of access to typical university staff. Video conferencing was inconvenient and rarely used, and voice-over internet did not yet exist. With communications at the time vastly more problematic, one can look back, again, and feel some surprise as to how the project could have advanced and converged at the speed it did. Even now I find it a difficult question to answer. Probably there was greater thought given to the necessity of each piece of communication, and more care in copying information to those who needed to see it. Decisions were, perhaps, more carefully considered, and generally more binding.

I STILL LOOK BACK WITH SOME SURPRISE that all of the necessary scientific and technical elements did, indeed, come together to meet the results that had been adopted. In the late 1970s, an accuracy of two milli-seconds of arc was targeted and finally accepted by ESA's advisory committees. Of course, much thought and analysis had gone into the early studies. And yet a great number of technical barriers seem to have come together, almost as if by chance, to allow this to happen. The beam-combining mirror was at the very limit of industry's manufacturing ability. So too were the detectors. The data rate sent down was at the limit of transponder technology, and we ran at the edge of on-board processing power, solar array technology, and a whole host of other support services. We operated at the limits of ground-based computers and data storage.

There were scientific providences or coincidences too. Corrections for the aberration of starlight demanded a state-of-the-art understanding of the Earth's motion around the Sun, available just—but only just—to the accuracy which was needed.

Correcting for the effects of gravitational light bending by the Sun demanded knowledge of general relativity, and we would not have been able to assemble the celestial jigsaw had this theory not been discovered more than half a century before. Our pieces would, quite simply, not have fitted together. Whether this would have been a disaster for the project, or in contrast, its greatest triumph, I cannot judge. Would just one brilliant mind have been able to connect the enormous discrepancies with the moving position of the Sun, and dared to have suggested that Newton was wrong?

THE SIZE AND DURATION of the scientific collaboration merits a specific mention. Lennart Lindegren's (originally Erik Høg's) consortium numbered nineteen scientists. Jean Kovalevsky's was larger at eighty two, of which thirty were French and the same number Italian. Erik Høg's Tycho team totaled thirty five, although a number were already active in the other teams. Catherine Turon's input catalogue effort counted twenty one major players and thirty three other contributors, numerically dominated by French scientists. Adriaan Blaauw's selection committee comprised fifteen senior scientists although their effort, while crucial, was much more limited in duration. The science team numbered sixteen, also drawn from the teams already counted. As the numbers make clear, many scientists were involved. Not all of these counted worked for the full duration of the effort, nor would they have worked on the project full time–their activities were invariably supplanted by other academic commitments, including university teaching.

The costs born by European science in support of these efforts is difficult to quantify, complicated by different charging policies and accounting procedures. Very roughly, the total involvement of nearly two hundred scientists within the four teams probably corresponded to the equivalent of about sixty full-time people per year averaged over the sixteen year lifetime of the project, of which the input catalogue preparations consumed, perhaps surprisingly, nearly half. To these would need to be added the appropriate computer, infrastructure, and travel costs. Computer usage, though fundamental and significant, was rarely billed to the project itself, more typically being part of the university infrastructure included as staff overheads.

OUR EFFORTS TO PUBLICISE THE HIPPARCOS RESULTS in the popular media were always a surprisingly difficult challenge. My own attempts at stereo projections aside, awkward to replicate for large audiences, the project produced no spectacular images. And turning the rather dry content of the star catalogue listings into something that would excite the average European taxpayer was a task demanding resources to address, and suitably talented individuals with time, experience and commitment to undertake.The many different languages of the various ESA member states compounds the efforts needed to penetrate the popular psyche across Europe. Leading science writer Nigel Calder contributed to a number of news releases and other stories which we crafted together, some of which proceeded smoothly, others not quite so. Once, sending off a text referring to the "grand Hipparcos catalogue of 10^5 stars", unwisely in retrospect using scientific notation in the submission to refer to our bounty of 100 000 suns, I was dismayed to see, in print, a reference to ESA's two-decade, multi-million high technology space survey which had apparently measured a total of a mere 105.

Chapter 10

LOOKING AT THE HISTORY OF SCIENCE over two millennia, it is clear that advances in intellectual thought or experimental design have all-too-often been overtaken by setbacks in the structure of societies that put an end to further progress for centuries. As Herbert Fisher phrased it in his *History of Europe*, first published in 1936: "The fact of progress is written plain and large on the page of history; but progress is not a law of nature. The ground gained by one generation may be lost by the next." It is tempting to imagine that this is a relic of past history, and of irrelevance to society today.

There are, however, many complex circumstances which must unite to make such grand undertakings possible. Amongst them are the willingness of governments to fund research, and universities that encourage it. Progress demands academic freedom to pursue knowledge for its own sake, and the stimulus of competition and prestige, albeit modulated by the desire to collaborate and share knowledge and ideas. It needs scientific journals in which new results can be published authoritatively, acting as a catalyst for new ideas, as well as libraries and computers that hold our vast stores of knowledge, and which allow for its immediate access around the world.

Progress needs industries with ever-advancing technologies; intergovernmental agreements that permit organisations like ESA to exist, fostering the advance of ideas and experiments well beyond the capabilities of individuals, institutes, or even individual nations, and providing structures that allow large projects to be managed through to successful completion. It needs a synchronised advance of ideas and technology, culminating in bold experiments which, while state-of-the-art, somehow still remain feasible. Advances also rely on the enthusiasm of an inspired public prepared to see some of society's wealth focused in this way.

Towering above all else, progress rests on the indomitable spirit to pursue some vision with boundless energy, to devote entire lifetimes to a single task, and to overcome all obstacles in an attempt to prove the human brain up to the task of decoding the riddles that Nature has thrown at our feet.

I WILL LEAVE IT TO OTHERS, more inclined to speculate, to reflect on whether we remain on a continuous wave of advances which will grow still further in the future. Or whether, more akin to the transient peaks of the classical Greek and Roman civilisations, we have lived through the current Golden Age, in which too many obstacles to future collective progress now lie ahead. In a finite world with finite resources, is indefinite growth assured?

Chapter 11

Our Galaxy

We find that we live on an insignificant planet of a humdrum star lost in a galaxy tucked away in some forgotten corner of a Universe in which there are far more galaxies than people.

Carl Sagan (1934–1996)

W HEN THE HIPPARCOS CATALOGUE dropped into the laps of the world's astronomers in 1997, it painted a fundamentally new picture of the way in which the stars are arranged and move through space. In these final three chapters, I will give some sense of how our understanding of the Universe has developed as a result.

When professional astronomers describe our Galaxy, it must appear terribly confusing. The few thousand stars visible to the unaided eye, smattered haphazardly across the sky, give few clues about the reality of its flattened disk, its vast spiral arms and its great star clusters. Telescopes are needed to take us outwards into the millions and billions of stars that are assembled into the various structures together comprising its profound complexities.

The great spiral galaxy Messier 31, or M 31[43], two million light-years away in the constellation of Andromeda, is not a bad proxy for how our own Milky Way galaxy might appear if viewed from afar. It is the nearest spiral galaxy beyond our own. Well-defined, clumpy spiral arms are visible sites of active and ongoing star formation. On such a panorama, our own Sun and solar system would lie far out in one of these magnificent spiral arms. It's useful to keep the image in mind when looking at the features of our own in some more detail.

M. Perryman, *The Making of History's Greatest Star Map*, Astronomers' Universe
DOI 10.1007/978-3-642-11602-5_12, © Springer-Verlag Berlin Heidelberg 2010

GALEX ultraviolet image, NASA

The great galaxy in Andromeda

Because images of our own night sky, or of these distant galaxies, appear unchanged over decades and even down the centuries, it can be difficult to grasp that our Galaxy, and others, are in a state of perpetual rotation. On top of this general rotation, the stars are flying around in complex patterns and at enormous speeds, dying, exploding, and being recreated again in an all-but-endless self-sustaining loop. Space is so very large and the timescales for these astronomical phenomena are so very long that it's easy to identify with Flann O'Brien's eccentric protagonist in his surrealistic novel *The Third Policeman*. When confronted with old cinematographic films, de Selby described them as having a strong repetitive element, and as being tedious; he had "examined them patiently picture by picture and imagined that they would be screened in the same way, failing at the time to grasp the principle of the cinematograph." In a similar way, photographs of the heavens inevitably convey the impression of a static unchanging universe arranged around us.

Nothing could be further from the truth. To get us off on the correct track, it must be fixed firmly in mind that the stars within our Galaxy are in constant and complex motion, their patterns changing continuously, erupting from clouds of gas as if from nowhere, to live and to die, diffusing and merging back into the disk of the Galaxy. The disk itself is in perpetual rotation, a giant celestial wheel, some stars like the Sun moving in more-or-less circular paths, others plunging around it in hugely eccentric orbits, moving at unimaginable speeds. All are dipping up and down across its central mid-plane, while here and there groups of a thousand or more are moving in tight formation as star clusters. Far out, in a vastly more rarified spherical halo, very old stars move high above the plane and occasionally plunge right through it. These strange travelers are rare and ancient remnants of our Galaxy's formation billions of years ago.

In thinking of our Galaxy, we must not picture a static population of stars. With our perception of time speeded up by a factor of a billion, it would be, rather, a choreographed performance of staggering complexity. The superficial impression of stability and tranquility, and the actuality of our Galaxy in perpetual turmoil, are connected by the stupendous scales of time and space on which significant changes occur.

A SCREENING OF THE UNIVERSE'S thirteen billion year history compressed into a two-hour feature length film would help to set the scene. Our Galaxy would only start to appear half an hour from the start, gradually taking shape through the settling mists of primordial hot gas as it cools and collapses into vast dark matter wells. Imprints left by the Big Bang, these gravity traps demarcate the embryonic structures that will in due course become galaxies. Swirling eddies impart rotation to the nascent structures, which gradually flatten and become more well-defined. Stars are formed from the gas clouds collapsing within the outlines of each galaxy. Turning even as it is born, each grand rotation of our own Galaxy, two hundred and fifty million years in cosmic time, is thereafter played out every two minutes of Spielberg time.

As the film progresses, the soundtrack silently mirroring the vacuum of space, stars are born in groups of a few hundred. These are embryonic star clusters, expanding outwards in a few seconds and scattering quickly through the disk. The most massive stars within them die out a few seconds after their on-screen birth, while those much less massive than the Sun smoulder away slowly, almost unchanged even by the end of the film.

Panning to the outer reaches of the Galaxy, and zooming in a billion times, our Sun itself—and its attendant system of planets, satellites and other debris—appears on the scene just over half way through. From that pivotal point on, a year of Earth time flashes by one million times a second. The total duration of *homo sapiens* on Earth, one hundred thousand years of central importance to our audience, occupies in its entirety just the final frame, a mere fraction of a second before the credits roll.

A snapshot at the end of our film would reveal our Galaxy, just as in the image of Andromeda, to be a highly-flattened disk of billions of stars, arranged in a few conspicuous spiral arms, with the Sun in the far outskirts. The Milky Way arching across our own night sky is just the star-filled disk of our Galaxy in which our Sun is immersed. It is at its densest as it passes through the constellation of Sagittarius, and this signposts the direction of the centre of our Galaxy. Our sight line through the disk crosses the largest numbers of stars in this direction, while a hundred and eighty degrees away on the other side of the sky, we peer outwards into the farthest expanses of our Galaxy, and onwards into the fathomless depths of intergalactic space. Within the endlessly rotating disk, spiral arms mark out sites of intensive star formation, and resulting concentrations of hot young stars.

Are we alone in this shocking vastness, racing about our daily lives largely oblivious to the immensity of our Galaxy and the Universe in which it is immersed? Or does our Galaxy, and its billions of stars, teem with life, a dazzling diversity of forms far beyond the imagination of even our most extreme science fiction exponents?

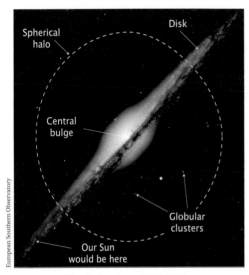

The disk galaxy NGC 4565:
how our own might appear edge-on

In the centre of our Galaxy, thousands of light years distance from us, is a vast spherical bulge of stars which dominates the light of the central region. More careful inspection reveals that the whole highly luminous flattened disk, shining by the light of its billions of stars, is immersed in a much less populous and far less luminous spherical halo of old stars. Instead of moving in circular or elliptical paths in the disk, these halo stars are arranged in a more spherical pattern around it.

This basic structure was known already decades ago. Hipparcos has sharpened the picture, and improved the understanding of our Galaxy's present structure and past evolution.

THE DISTANCE FROM OUR SUN to the centre of our Galaxy is the most fundamental of all numbers entering into discussions of its structure and motion—as basic as the radius of the Earth, or the distance from the Earth to the Sun. Something around thirty thousand light years away, its enormous distance is still rather poorly known. Stars near the Galaxy's centre are far too distant for direct triangulation through the method of parallax, even with Hipparcos's remarkable space-based acuity. We have to resort to more indirect methods to estimate its vast distance.

The 'standard candle' technique in astronomy is a way of estimating distances using intelligent guesses about the properties of certain stars. It depends on identifying classes of stars which have a reasonably fixed intensity. If we can find such stars, and if they are both common and luminous, we can proceed as follows. If we see some of the class nearby to us in space, we can measure their luminosities directly. If we also see them at some far away location whose distance is needed, we can use their known luminosities and apparent brightnesses to infer their distances. The method is sprinkled with difficulties, but it's often that or nothing. It's a method especially difficult to use in the case of the Galactic centre, because large and patchy gas clouds piled up there make it difficult to correct for the attenuation of star light taking place within them. Various special 'standard candles' currently provide our best distance estimates.

Extraordinarily bright and highly-pulsating stars called Cepheid variables have been used to stake out distances far across our Galaxy. More than two hundred in our region of the disk have had their space velocities measured by Hipparcos. They share a rotational pattern which gives a strong clue as to the distance of their centre of rotation. RR Lyrae variables are another standard ruler which can be used in much the same way. Hipparcos also measured hundreds of helium burning giant stars in our solar neighbourhood, and they too provide a luminosity calibration which allows others much further into the Galactic centre to have their distances estimated.

There is a totally independent method which makes use of a very special creature at the centre of our Galaxy—our very own black hole. Deep in the constellation of Sagittarius, a bright and very compact source of radio emission appears to pinpoint a region of gas and dust which is being heated to millions of degrees by some extremely bizarre phenomenon. The arguments now stack up that what we are seeing is material falling into the irreversible grip of a massive black hole. Infrared maps can now trace the paths of stars encircling the region, and their orbits confirm that some immense gravity 'well' must reside there. These are the manifestation of an enormously weighty and extremely compact object, a mass of more than three million suns packed into a region smaller than the orbit of Mercury. We can get a distance estimate to the black hole from the scale of the orbits encircling it. All these methods tell us that the distance from the Sun to the centre of our Galaxy is around eight thousand parsecs, or some thirty thousand light years—light which left the centre of our Galaxy thirty thousand years ago is only now just reaching us.

THE WAY IN WHICH OUR GALAXY ROTATES about this distant centre remains something of a puzzle. In the early years of the twentieth century, Jacobus Kapteyn put forward a model for our Galaxy based on star numbers counted from the first great photographic surveys. Almost echoing the ancient geocentric view of the Universe, he inferred that the Sun was close to its midpoint. He believed that the density of stars fell away with distance from the centre, defining a flattened system with a radius of about four thousand parsecs, more than ten thousand light-years. Interest in the shape and size of the Galaxy precipitated astronomy's 'Great Debate' in 1920, and studies have been pursued relentlessly since.

Understanding made a great leap forward with the work of Bertil Lindblad (1895–1965) in Sweden, who had been working to understand the rotation of the Galaxy. In an influential paper in 1927, Lindblad proposed a radically different model in which the entire Galaxy rotated around a far more distant centre.

Per Olof Lindblad

Bertil Lindblad

Lindblad was also puzzling over the nature of our Galaxy's spiral arms. Grand spiral galaxies, like our own Milky Way, are two-a-penny in the Universe. We now know that their magnificent overriding features, their majestic spiral arms, mark out the sites of intensive star formation. In 1925 Lindblad realised that the idea of stars arranged for perpetuity in a spiral shape was not tenable. He already knew that the speed of rotation of the Galactic disk varies with distance from the centre, and this led to a conundrum. After a few rotations, the arms would become increasingly, and implausibly, more tightly wrapped.

Lindblad proposed one of the two models that now hold sway to explain them. In this picture, the spiral arms are regions of higher density pressure 'waves' that—for some reason—are rotating more slowly than the galaxy's stars and gas. As any left-over gas in the disk catches up with one of these density waves, it becomes compressed as the wave passes by. Squeezed such that it collapses in upon itself, becoming denser and hotter in the process, stars burst into life once the densities reach the critical state for nuclear fusion to take hold, and a new cycle of star formation starts again.

Young star associations which we see around us are the sites of recent star formation which have appeared within this compressed gas, and it is mainly massive blue stars that light up these spiral arms. Very massive stars are born and burn through their fuel in a few million years, in such a short instant of cosmic time that they appear and disappear again almost instantaneously as the density wave passes through. Less massive stars, like our Sun, will burn on for hundreds of millions or billions of years, and will continue shining long after the density wave has moved on, adding to the stars already marching on their long journey around the rotating Galaxy.

Seventy years after Lindblad's pioneering work, and it is still not particularly clear how these ubiquitous density waves originate, how fast they propagate, or how long they last. They appear on the scene like whirlwinds out of nowhere, compressing gas in the disk as they advance around the rotating Galaxy. It may be that they persist for one or two Galactic years, a couple of rotations lasting half a billion years, and then fade out to be replaced by others. Or they may continue to trundle around the Galaxy for several turns. Explicable or not, these star formation bursts do start up in various places in our Galaxy, and new stars appear out of the residual gas. They are left shining for millions or billions of years, dramatic evidence of huge forces at play, drawing attention to their birth pangs long after their life-giving forces have moved on to work their renewed magic in pastures new.

Jan Hendrik Oort (1900–1992) from The Netherlands entered the discussions of the Galaxy's rotation shortly after Lindblad's suggestion that our Sun lay far distant from its centre. Having studied under Kapteyn in Groningen in The Netherlands, Oort had crossed the Atlantic to cut his astrometric teeth measuring photographic plates at the Yale observatory in 1922. He then returned to Leiden, where cosmologist Willem de Sitter and star classifier Ejnar Hertzsprung were successive directors, to continue his studies of a puzzling class of high-velocity stars.

Jan Hendrik Oort

Oort showed how Bertil Lindblad's idea of rotation about a distant centre was supported by the observations at his disposal. He introduced a new description of the Galaxy's rotation, neatly solved the problem of the high-velocity stars, and estimated both the local rotation speed and a distance to its centre in the process. Many adjustments have been made since, those of Hipparcos being the most recent and discerning. While Oort's estimate of about 5900 parsecs is quite a way from our present figure of 8200 parsecs, it was an impressive result nonetheless. It established Oort's name amongst the role call of influential astronomers. In the early 1950s, his investigations into the nature of the cometary cloud surrounding the solar system which now also bears his name, propelled him to wider renown—in 1955 *Life* magazine included him amongst its hundred most famous, alongside Eisenhower, Churchill, and Khrushchev, rubbing shoulders with Stravinsky and Picasso.

Jan Oort remained an influential figure in twentieth century astronomy; an obituary lamented that "The great oak of astronomy has been felled, and we are lost without its shadow." He continued a close involvement with Leiden Observatory well beyond his formal retirement in 1970 until his death in 1992[44]. Our paths overlapped there with his own interest in what Hipparcos would soon have to say about our Galaxy's structure.

THE OORT–LINDBLAD MODEL is the name now bestowed on a simplified picture of our rotating Galaxy, in which the disk is assumed to be turning uniformly about a central axis. Astronomers use the label 'Oort constants' for two important numbers which describe the large-scale motion which is observed. This rotation was first measured from the line-of-sight motions of stars at different directions in the disk looking outwards from the Earth. Over the decades since, many attempts have been made to characterise our Galaxy's rotation.

The spiral galaxy M81, seen in a composite of ultraviolet, visible and infrared light

There are lots of reasons for wanting to know the details better. We'd like to know the speed of the Sun's rotation around the Galaxy, and to establish whether the Sun is moving along with most of the other stars or in some different way. It may appear surprising, but this helps in understanding questions related to long-term habitability on Earth. We'd like to know about the up-and-down motion of the Sun through the plane of the Galaxy, because there seem to be long-term climate variations on Earth, and bursts in meteorite cratering records, that might be explained by this long term vertical oscillation.

The rotation speed is controlled by the total mass making up the entire Galaxy, which in turn dictates how the Galaxy was formed during the early phases of the life of the Universe. We'd like to know how the rotation speed changes with distance from the Galaxy centre, and we'd like to know the fraction of the total mass that's hidden in the form of dark matter—both would help to establish whether our present cosmological models are correct.

We'd like to know how the ghostly spiral arms rotate. More pressingly, we'd like to understand whether the Sun is moving around the Galaxy at the same speed, or faster, or slower, than the moving pattern of the ethereal spiral arms. This is because, over many tens of millions of years, our Sun's past passages through the arms must have affected the environment in the solar neighbourhood and, in its turn, the long-term climate on Earth.

THE OBSERVATIONS THAT WE NEED in order to understand the rotation can be specified quite easily, but they're difficult to make in practice. Measuring the space motions of just one or two stars is no help at all—in addition to the strong overall pattern in the rotational motion, there is also an enormous amount of chaotic or random motion sprinkled on top. Some stars have very large space motions, perhaps the result of some specific encounter with another star in the distant past. Naturally, if we happened to measure that one, or that one as part of a small sample, we'd have a distorted picture. And we now know that the youngest stars move through space in a very different way to the oldest. Those just born received a type of compression shock which made them collapse into stars in the first place, while the oldest have, over the hundreds of millions of years since their creation, settled into a more stable equilibrium.

A further complication is that some of the stars traveling through space around us are not part of this rotating disk at all, but part of a very different structure—the Galaxy halo—in which the disk itself is immersed. It's all very messy. Our challenge is almost like sitting on a life-raft somewhere in the inner reaches of a massive bubbling and churning whirlpool, trying to estimate where we are by taking the occasional sightings of land.

Figuring out these complex motions from our location buried inside the Milky Way is harder than seeing the overall picture from a distance, even though we are right up close to the action, surrounded by important clues. From high above the great orbital motorways—the M25 around London or the Périphérique in Paris—we would be able to see the average flow rate, and could easily estimate the total distance around the city, or the average speed required for a full circuit. But from a car somewhere in one of the busy lanes—no map or GPS to serve as guide—it would be a challenge to figure out a precise location or a meaningful average speed.

By measuring distances to luminous stars right across the Galaxy, and the motions of dwarf and giant stars, hot young stars, and stars in spiral arms, the picture is now much clearer. At the same time, our new precise measurements reveal the bewildering complexity of stars in motion around us. Near to the Sun our region of the Galaxy is turning slowly, the Sun more-or-less with it. But there are sheering and drifting motions as well, and pockets of stars expanding swiftly outwards from the regions in which they formed a few million years ago. Out of all this turmoil, we can distill some order.

OUR BEST ESTIMATE of our Galaxy's rotation show that, in our neck of the woods, it is moving, en masse, at a speed of about 223 kilometers per second. As a result, our Sun—and our solar system with it—makes a complete circle of the Milky Way once every 226 million years. Like the yearly orbit of Earth around the Sun, this is our new measurement of one Galactic year.

These speeds merit a closer look. Newton's laws of gravity, cemented by Einstein's relativity, caution us that we can't be aware of our velocity as long as it's constant. It's a view we might accept when riding in a smoothly ascending lift, high-speed train or high-flying plane. But is it not an affront to our intuition to be asked to believe that, with the Earth turning once a day on its axis, we're actually spinning through space at a dizzying five hundred meters per second, or a little short of two thousand kilometers an hour? Worse still, that by clinging to the Earth's surface in its annual sweep around the Sun, our additional speed as we go around it is a remarkable thirty kilometers *per second*, or a heady one hundred thousand kilometers per hour. At this thundering pace, we travel a princely two and a half million kilometers around our Sun each day.

Yet these alarming speeds and huge distances are as nothing compared with those we travel as our Sun, dragging its solar system with it, participates in its rotation around the Galaxy. At the speed of a little more than two hundred kilometers per second clocked by Hipparcos, just short of one million kilometers per hour, we journey twenty million kilometers each day through the vastness of space. Wonder at the figures, feel awed by the numbers, and take a look out of the window to double-check your senses. But just as if challenged by the police for speeding, denying it through lack of awareness won't change a thing. Intuition is often an unreliable guide in the mountainous landscape of science. Werner Heisenberg, the gregarious Bavarian intellect behind quantum mechanics, offered his own conviction that in physics it is a basic mistake to rely upon it.

IN RECENT YEARS it has become increasingly clear that the solar system is a very special place for all sorts of reasons, and its velocity through space is one of them.

Our Sun doesn't quite participate in the average motion of all the nearby stars in this mob rotation around the Galaxy. Astronomers refer to the average motion of the stars around us as the 'local standard of rest', a priceless understatement for such a reckless speed. Watching the stars moving around us, its sight bearings unmatched by celestial surveyors before, Hipparcos tells us that we have a small additional motion on top of the average procession: an extra few kilometers a second in the direction towards the centre of our Galaxy, a few more in the direction of our rotation around it, and a little extra up out of its central plane too. In total, about thirteen kilometers a second above and beyond our local standard of rest—but what's the odd fifty thousand kilometers an hour, here or there?

Not so much, fortunately as it turns out. The behaviour of the stars is much akin to the motions of atoms in a gas, where all have their own random motion. The erratic part of the Sun's dash through space is quite small compared with all the motions going on around us. This is good news, since a small local speed, just as for a car on the roads, lessens the likelihood of a crash with another traveler.

As we gallop through space at these enormous speeds, is a crash with another star on the cards? Such a collision would be serious indeed, but we can figure out the chances of a head-on impact given our speed and the local traffic density. It turns out that, even though the speeds are high, space is so very very empty that we can expect to hit, or be hit, only once every ten thousand billion years, give or take. A direct hit remains improbable even after the passage of time equivalent to a thousand times the present age of the Universe.

Even a near miss, though, could be disastrous. Again, the chances are also considerably reduced at our low speed, especially during the Sun's motion up and down through the mid-plane of our Galaxy. That the Sun moves up and down through the plane is a property that we share with every other star. It's just a consequence of the gravitational pull of all the other stars spread out in the Galaxy disk that pulls each star back down each time it tries to struggle free.

OUR SMALL UP-AND-DOWN SPEED, measured precisely with Hipparcos, means that we will not rise far above or below the plane before gravity pulls us back down to shoot through to the other side, before being pulled back again, and so on, oscillating this way for the rest of eternity. This means that we do not show our heads far above the parapet of the protective bath of hydrogen gas which permeates the mid-plane. If we did, we would be an easy target for the X-rays and gamma-rays pervading this emptier region of space. If our up-and-down motion had been more pronounced, life would not have been cradled to the extent that it has over the billions of years since it first struggled to take hold on Earth. We owe our existence to the small motion of the Sun compared to the nearby stars.

Neither will the small component of velocity towards the Galaxy centre take us far towards its inner reaches. The course we steer around the Galaxy will keep us clear of the high star density at its centre, at least for a very long time. Here again, our Sun's path is highly felicitous. Whether this is a remarkable happenstance, or a prerequisite for the development of life on Earth, we're not yet clear.

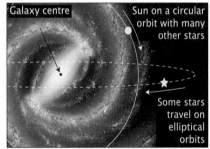

Simulated face-on view of our Galaxy

Last but not least, the very small component of extra speed in the overall direction of our Galaxy's rotation suggests that we are not far from rotating in step with monstrous density waves which are propagating around it, the shock waves responsible for the Galaxy's prominent spiral arms. As a result, these mysterious density waves will make their ghostly passes over us only rarely. When they do, the waves will compress the gas and dust in the surrounding region, giving rise to new bursts of star formation around us. This is not such a problem in itself. But in only a short cosmic time following a burst of star formation, the most massive stars will die off as supernovae, hurling off their outer mantles, and eliminating anything in their vicinity with a hideous blast of X-rays and gamma-rays. Frequent spiral arm passages would mean frequent gamma-ray blasts, and this might be bad for life on Earth too.

FROM OUR VANTAGE POINT encircling the Sun, we can peer out and see that our solar system occupies a position rather close to the plane of our Galaxy's disk—close to its mid-plane, but not exactly so. The disk is around a hundred parsecs in width, three hundred light-years from top to bottom. Stars of all types are reasonably smoothly distributed over this layer, although many sit out at much larger distances. We can count the number of stars out to different distances all around us, and this three-dimensional mapping can be used to judge how far we are from the central plane. The answer is that we are about twenty parsecs, about seventy light years, from the mid section. Our Sun pokes out just above the crowded plane, and therefore gives us a slightly better view of our Galaxy all around us than if we'd been more deeply immersed within it.

The precise distance from the mid-plane is really of only secondary interest—more so is the extent to which the Sun bobs up and down above and below it, and the period of this bobbing motion.

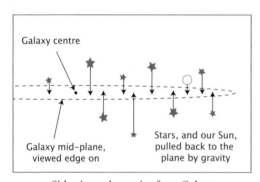

Side view schematic of our Galaxy

We can liken the stars spread throughout the disk as being like corks on the surface of a sea. If the sea is very calm, all the corks would spread out in a thin layer. If the sea becomes more choppy, the corks bob up and down, the vertical spread becoming larger as the sea becomes rougher. A snapshot would show some corks high above the average sea level, others below, and some around the midpoint on their way up or down. If, at the same time as making the snapshot we could estimate their speeds, up and down, we could figure out a lot about the properties of the waves: their typical height, and their frequency being the most important.

Stars in the Galaxy plane behave in much the same way—they also bob up and down, with a period of oscillation dictated by the total amount of mass in the disk. Each individual star is pulled back down as it rises away from the mid-plane, by the collective gravity field of all the other stars—even though they are also bobbing up and down individually, the average centre of mass stays fixed in the middle. Measure the distance to the stars, and estimate their masses, and we can figure out how much mass is tied up in the disk stars. Measure their vertical velocities in addition, and we can estimate the total mass tied up in the disk in whatever form—stars, gas, and even dark matter, as well as the duration of their up-and-down motion.

The Hipparcos results have given some clear answers to these difficult questions. The mass in the disk is about one tenth of the mass of the Sun in each cube of space one parsec on a side. In other words, if this matter could all be spread out uniformally across the mid-plane of the Galaxy, it would have the thickness of a sheet of normal writing paper. It may not seem particularly heavy, but you would struggle to pick up a cubic meter of paper—it weighs in at more than a ton. Spread across the vast expanse of our Galaxy, it adds up to several billion times the mass of our Sun. It is this mass which determines the up-and-down motion of stars in the disk, by fixing the local gravitational force of all the matter combined. The Hipparcos studies pin down the period of oscillation, the bobbing motion of stars up and down through the disk, to be around eighty million years. Our Sun itself, and our solar system with it, are not immune to these forces, and we we are also oscillating up and down, through the disk, once every eighty million years.

There are interesting consequences. The mass which we infer to be controlling the motion is actually very close to the total amount in the neighbourhood of the Sun that can be seen around us—it matches the total once we add up the mass in stars, white dwarfs, gas, and interstellar 'dust'. This means, in turn, that all the mass that can possibly be in the vicinity of the Sun has been properly accounted for. Although this may not sound too surprising, it's a great comfort in science to know that the book-keeping has been done correctly. It has, moreover, an important consequence for present-day discussions in cosmology about the nature of the invisible dark matter—for whatever dark matter is, and wherever it is hiding, we now know that it's not distributed just like the disk itself, in some highly flattened form. It must be spread out, and occupying a much larger volume, probably arranged in the shape of our Galaxy's halo.

JUST AS IN ANY CENSUS, collecting the data is one thing, making sense of it is quite another. To divine more about the nature of the stars, more information than just the distances and space motions is frequently needed. Many surveys of the skies before and after Hipparcos—for example, in infrared or X-ray light—contribute considerable extra knowledge.

Classifying the chemical make-up of the enormous numbers of different stars is an endlessly challenging task which has been ongoing for decades, and which has progressed rapidly in recent years. Bigger telescopes have made the light collection more efficient, and better spectrographs have made the hieroglyphics encoding the elemental abundances much more easy to read. Astronomers can now turn to their centralised data banks to pluck out the elemental abundances of the stars contained in the Hipparcos catalogue. With the space-based mapping, it suddenly became much easier to study their chemical make-up according to their position in the Galaxy, their space velocities, their height above the plane, or their connection with sites of star formation.

Though one star might look like any other to the non-expert, taxonomists have been successful in dividing up the stars into various significant categories. Like classifying people into young or old, tall or short, dark hair or light, groupings may overlap. Similarly, stars are classified according to their temperature, segregated into giants and dwarfs depending on their size and luminosity, differentiated into young and old, and categorised as those primordial in their chemical composition or those enriched by heavy elements as a result of nuclear burning. A few short decades before Hipparcos another important classification was recognised. Stars in our solar neighbourhood are partitioned into two important classes: disk stars and halo stars, although the former outnumber the latter by a factor of a hundred.

We have met the disk stars already: like the Sun, they rotate around the centre of the Galactic, pretty much confined to its plane. The disk stars can be divided further into young and old, and into dwarfs and giants, each with their own mass and luminosity. But in terms of how they bounce up and down around the plane of the disk in the course of their journey around the Galaxy, there are two subtly different sorts: those with relatively little energy (like our Sun) which don't bounce too far (astronomers call them the thin disk), and others with lots of energy which bounce energetically to positions high above and far below the plane before falling back down (the thick disk). Hipparcos gives the distinction on a star-by-star basis, assigning a space motion and an age to each, essentially by looking up its properties in a handbook of stellar models.

The result is that we can now assert that the main 'thin' disk of our Galaxy came into being early on in the life of the Universe. The thick disk stars were added into the mix billions of year later, probably as our Galaxy captured and devoured an unfortunate interloper which strayed too close.

THE HALO STARS are a still more curious population. Their name is descriptive. In addition to the billions upon billions of stars which together comprise our Galaxy's flattened disk, and which rotate together like a giant pancake turning every quarter of a billion years, the disk is encased within an almost spherical halo of stars.

Several properties mark this halo population out as being very special: there are not many stars making it up, they do not rotate in step with the disk, and they are all very old. Our Galaxy's halo is a vast, spherical, and thinly populated distribution of ancient stars which surrounds the dense rotating disk deep within it. But what is it, and where did it come from?

We must step back two centuries to the work of the British astronomer Stephen Groombridge (1755–1832). Using a transit circle, Groombridge compiled a star catalogue to about eighth magnitude, a little fainter than those visible by eye. For his careful measurement work leading to this *Catalogue of Circumpolar Stars*, published in 1838, he was duly elected as a Fellow of the Royal Society. Few professional astronomers today would know the details of his labours, except for one of the stars which still carries his name and catalogue number—the sixth magnitude star Groombridge 1830. It seemed unremarkable enough, but charting its position would leave its legacy.

Prussian astronomer Friedrich Argelander (1799–1875) was also busily devoting his efforts to measuring star positions, and the pinnacle of his life's work was the massive Bonner Durchmusterung survey of more than 300 000 stars. The observations of Groombridge, and the later survey by Argelander, were spaced a couple of decades apart. From the comparisons of any two catalogues, individual star positions can be checked. While most stars, as we have seen, stay pretty much fixed over decades or move only slightly, rarities with particularly large motions can occasionally be picked out.

What emerged from the comparison of these two catalogues was a real gem—the star with the highest motion ever seen, Groombridge 1830. It was announced by Argelander in 1842. Its discovery nudged the previous record holder, 61 Cygni or Piazzi's Flying Star, into second place. While for most stars in the heavens a typical annual proper motion is about three hundredths of a second of arc, Groombridge 1830 is fairly zipping across the sky at slightly more than seven seconds of arc each year. Over a hundred years, it moves by one third the diameter of the full Moon. As record holder it was in turn replaced by the fainter Kapteyn's Star discovered by Jacobus Kapteyn in 1897 with a motion of more than eight seconds of arc a year,

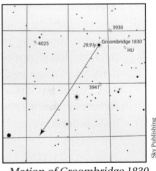

Motion of Groombridge 1830 in the Millennium Star Atlas

and that by the present record holder, the slightly fainter Barnard's Star discovered by E.E. Barnard in 1916 at just over ten seconds of arc a year.

Although news of these still tiny motions is unlikely to result in sleepless nights for the masses, they cause more than a shiver of excitement amongst the professionals. All three of these faint fast-moving objects are very close to the Sun, amongst the very closest. They are all somewhat less massive and also less luminous than the Sun. But most importantly, they are immensely stunningly old, predating our Sun by far. Barnard's Star is amongst the oldest objects in the Universe, and has been shining more or less at its present rate for more than ten billion years.

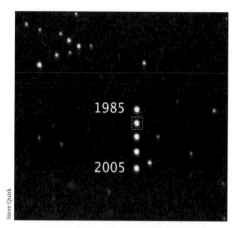

Barnard's star, from 1985 to 2005

Steve Quirk

Barnard's Star shot to prominence in the early 1960s following a claim by Piet van der Kamp (1901–1995), a Dutch astronomer working in America, that he had detected a wobbling motion in its position on the sky. He had been observing the star closely since 1938, and duly announced that the wobble hinted at the presence of a planet circling around it. Claimed as the discovery of the first extra-solar planet, it sparked off a dispute which remained unsettled for decades.

As one of the nearest stars to the Sun, and with the possibility of a planet in orbit around it, the star's fame amongst the science fiction community grew. Between 1973–78 it was adopted as a target for Project Daedalus: a design by the British Interplanetary Society for a fusion-powered rocket, aspiring to reach more than one tenth of the speed of light, and to gain Barnard's Star within fifty years. In the USA almost twenty years before, Project Orion had also conducted an engineering design study of a spacecraft powered by nuclear-pulse propulsion. I mention these studies because the intervening decades of engineering and science advance still place even the nearest stars far beyond any present capability to contemplate a visit. The Hipparcos observations, incidentally, rule out a planet of anything like the size suggested by Van der Kamp.

THE ASTROPHYSICAL INTEREST of Groombridge 1830 came to the surface in the 1950s, two hundred years after its discovery, with the realisation that it typified a new class of object: high-velocity, low-metallicity stars in very elliptical or heavily inclined orbits, hurtling through the solar neighbourhood. Rather than rotating in the plane of the disk stars, these halo stars are moving through a much larger spherical volume, symmetrically arranged about the Galactic centre. The periods of their long lonely orbits through the Galaxy can be billions of years. If they should happen to pass close to the Sun, within a few light years, they do so with a very high velocity, and with a speed and direction very different from those of the disk stars. Although they appear to move fast, it is actually the Sun which is moving swiftly as part of the overall disk rotation; the halo stars, such as Groombridge 1830, are effectively standing still with respect to the rotating disk—they simply appear to be moving, in a retrograde direction with respect to the disk, at very high speed.

In the 1960s and 1970s, interest shifted to how the halo might have formed. The disk came into existence through the relatively rapid collapse of a rotating, galaxy-sized gas cloud, billions of years ago. In contrast, two different theories were put forward to explain the formation of the halo: in one, the halo stars were believed to form from gas in free-fall from the outer parts of the Galaxy heading towards the plane of the disk. Free-fall just means that the star's motion is controlled by the gravity field of the disk. Although the term conveys the impression of urgency, falling from these colossal heights would still take around two hundred million years.

In the other picture, the halo was posited as being built up over billions of years from galaxies very much smaller than our own becoming trapped in the gravity field of the Milky Way, and torn apart as they began spiraling in under its non-negotiable gravitational embrace.

Simulation of galaxy disintegration leaving its trail in our Galaxy's halo

Both of these theories grew out of the study of just a few dozen of the very rare high velocity stars known in our solar neighbourhood, typified by the speedy but otherwise self-effacing Groombridge 1830. The importance of these almost primordial objects, known as subdwarfs, was such that great attention was given to including them in the Hipparcos observing programme. When the results of the space census was published, astronomers working on the origin of the halo immediately and eagerly devoured its content.

An important question is how, billions of years after the event, can astronomy hope to say anything authoritative about the halo's formation. The answer is that these stars are astronomy's very own fossil records. To be of value as probes, these fossils need to have lain largely undisturbed for the billions of years since their formation. That this is the case results from two notable features. First, their hydrogen burning is so slow, because their mass is so low, that their stellar atmospheres preserve, more-or-less intact, a record of the quasi-primordial abundances out of which they formed. Accordingly, their chemical signatures turn out to be not only rather unique, being so old and rare, but also remarkably well preserved over almost the entire age of the otherwise restless Universe. Second, are their pristine orbits. In their extravagant and isolated paths through the halo, which may take many billions of years to encircle, they have come into close contact with nothing. Only in this most recent passage through the Galactic plane, which will be over in the blink of an eye on the celestial timepiece, do they come close enough for us to catch a glimpse. We must try, with urgency, to read the Rosetta Stone inscribed upon them as they race past.

Amina Helmi (2007)

Such was the challenge picked up by Amina Helmi from Leiden Observatory, and her colleagues, who published a remarkable discovery in the journal *Nature* in 1999. They had sifted through the hundred thousand stars from the catalogue to select those with a particular fingerprint identifying their age and chemistry. They found just ninety seven from the entire catalogue with the properties they were searching for so carefully. Then they calculated their orbits around the Galaxy, tracing their motions backwards over their enormous lifetimes. And they found just seven sharing the very same motions. Like DNA matching, and as confident in its use, they had found the ancient relics of a single small galaxy which had strayed too close. It had been torn apart under the immense gravity field of our own Galaxy, the debris still moving sedately around its outer regions in loose formation.

Although this death sentence had been imposed several billion years ago, the stars have continued to encircle the Galaxy as they spiral very slowly inwards. It was the first convincing example of a small group of stars whose common origin, and shared orbit, prove that the halo has been assembled from the merging and accretion of sub-galactic star clumps. Others soon confirmed the discovery, and a small number of additional independent star groupings were later found. Astronomers could collectively breath a sigh of relief that they had found unambiguous evidence of an ancient catastrophe. It was a discovery signaling a significant advance in understanding as to how our Galaxy had come into being billions of years ago.

ONE OTHER CLASS OF OBJECT parades through our Galaxy halo in a still more majestic if far rarer form—the globular clusters. These are exotic, beautiful, and very tight groups of many hundreds of thousands of stars. They look superficially like the much younger open star clusters but are, in reality, large and ancient assemblages of very old stars. Some of these most splendid groupings are visible through small telescopes, clusters like Messier 80 being popular targets for amateur astronomers.

The imposing and austere Omega Centauri is the grand master of them all. Poised in the constellation of Centaurus, it is one of the few that can be seen with the naked eye, the brightest and largest orbiting within our Galaxy. It was listed in Ptolemy's catalogue two millennia ago as a star, identified by Edmond Halley in 1677 as a nebula, and first classified as a globular cluster by John Herschel in the 1830s. Some 18 000 light-years from Earth, it contains millions of ancient stars, around twelve billion years old. Stars in the central region are packed so tightly together, a tenth of a light year apart, that many thousands of its suns would occupy the space between us and our nearest star.

Globular cluster Messier 80

Some globular clusters are so old that, before Hipparcos, a few challenged our belief in the origin of the Universe itself—they seemed too old to fit within the age limits derived from cosmology. Hipparcos changed the models, and our estimate of their ages, and brought them back in line with the chronology of the Universe. They are rare—our Galaxy contains only about a hundred and fifty in total—and far too distant for Hipparcos to have measured individual stars within them. Yet in a stroke we now have a reference framework in which their orbits around the Galaxy can finally be tracked with some clarity.

Some are moving through space like the rare halo stars, while others move like the disk stars. Traced backwards in time, their orbits have taken them close to the centre of the Galaxy, where its gravity field must have ripped away their more loosely-bound brethren. This tidal-stripping consigns globular clusters to an eternal wasting away as they age. If their death throes are now clearer, their origin is still not well understood—they may be the cores of ancient dwarf galaxies, once several hundred times their present sizes, which were torn apart and trapped by our Milky Way Galaxy, rather like the smaller groups of infalling halo stars.

FROM THESE AND OTHER explorations, we have a much clearer picture of our Galaxy's structure and evolution than we did a decade ago. Our space-age celestial mapping provides a new understanding of its rich complexity and strange formation history. We find relatively peaceful regions of our tumultuous Galaxy, and we appear to inhabit one. Such advances have flowed simply from a better charting of the positions of the stars.

Chapter **12**

Inside the Stars

I ask you to look both ways. For the road to a knowledge of the stars leads through the atom; and important knowledge of the atom has been reached through the stars.

Sir Arthur Eddington (1882–1944)

I T WOULD BE QUITE SOME CHALLENGE to look up at the night sky, and figure out the age of the Universe from the positions of the stars. Although this may sound somewhat more like astrology than astronomy, I hope that by the end of this chapter I will have explained, at least in outline, how this is done in practice.

Our Sun is the nearest of all stars in the sky by far. Not surprising, therefore, that it should be understood much better than any other. Its proximity means that it provides some of the best observations to pin down our understanding of the structure of stars and their evolution. Its mass, radius and energy output are all known to one part in a thousand. Accepted values today give a mass of more than a billion billion billion tonnes, or 1.989×10^{30} kilogrammes, determined from the size and period of our orbit around it. Its radius is pinned down from eclipses at just under seven hundred thousand kilometers, a hundred times that of Earth. Its power output is 385 million million megawatts, as measured by satellites in orbit high above the atmosphere. Its surface temperature is around 5500 °C, estimated from its spectrum in a way similar to knowing that a white-hot poker is hotter than one glowing red. All these numbers are simply too vast to be easily or meaningfully comprehended.

M. Perryman, *The Making of History's Greatest Star Map*, Astronomers' Universe
DOI 10.1007/978-3-642-11602-5_13, © Springer-Verlag Berlin Heidelberg 2010

The age of a star is reckoned from the time that nuclear reactions begin to dominate gravity collapse as its primary energy source. We can date this period in the Sun's pre-history surprisingly well, using radioactive decay products in meteorites which have fallen to Earth. For the Sun, this moment was 4.53 billion years ago.

Our Sun apart, almost all other stars in the sky are seen as just point sources of light. Except for some of the nearest and very largest, they are just too distant for their diameters to be discerned, even through the largest of telescopes, and even from space. Surprising, therefore, that we can deduce so much about their detailed nature.

It turns out that we can discern their internal structure and chemical composition mainly based on their surface temperature, their chemical composition, and their energy output. The first of these is estimated from a star's energy distribution with wavelength, the second from absorption lines in its spectrum, and the third from its brightness and distance. The first two are not the subject of this story, and we shall skip over the details. But the third underpins the Hipparcos satellite discoveries, and a few more words are in order.

When we focus our attention on a typical star in the sky, it may be just an average star at some 'typical' distance. But it could also be a very luminous star much further away, or a much dimmer star much closer to us in space. However, once we know the star's distance, we can use it, together with its brightness as seen from Earth, to calculate the star's energy output. It's no different from an artificial source of light, be it a domestic bulb, a car headlight, or a lighthouse beacon seen from a distance: only when we know the distance can we infer how bright the source actually is.

Before Hipparcos we had estimates of the energy output, or luminosity, of some of the nearest stars; from the satellite's results we have much better numbers, for stars considerably more distant, and for a remarkable hundred thousand of them all over the sky.

THE BIGGEST BREAKTHROUGH in comprehending the power source of the stars came from the understanding of nuclear fusion in the early decades of the twentieth century. Before then, their energy source was unknown. During the nineteenth century, it was thought that the stars were powered by gravitational energy released as they collapsed from massive embryonic clouds of gas. Once it was shown that this energy source could not possibly sustain the Sun over the billions of years of geological time laid bare in fossil records, a more long-lived source of fuel was sought.

For a time it was thought plausible that the gravitational impact energy from meteoritic rocks and dust continually raining down on its surface from the outer parts of the solar system might sustain it. Such an assumption entered the authoritative calculation of the Sun's age published in 1864 by William Thomson, Lord Kelvin (1824–1907): his calculations placed it at less than a hundred million years old.

Lord Kelvin

As the theories of the internal structure of atoms and elements advanced, it became clear that there was one energy source available in the Sun that had been overlooked—nuclear fusion. Fusion is the process by which multiple atomic particles join together to form a heavier nucleus, releasing colossal amounts of energy in the process. (In contrast, nuclear *fission* is a process where a heavy atomic nucleus, such as uranium, absorbs a free neutron and splits into two smaller nuclei. Fission also releases energy, and is the basis of nuclear power stations and most nuclear weapons.) Occurring naturally in stars, fusion has not yet been tamed on Earth, but it is the basis of ongoing experiments to design cheap electricity-producing power plants. ITER, the International Thermonuclear Experimental Reactor in Cadarache in southern France, is being developed to exploit this sustainable energy source, and arguably represents mankind's most ambitious techno-scientific megaproject of all time.

The fusion of two hydrogen atoms to form a helium atom is a process in which the hydrogen atoms forever lose their identity. Fusion releases a quantity of energy far beyond chemical reactions in which the elements themselves are only rearranged or recombined, never transmuted. And it was known from the size of the Sun, and from the almost total preponderance of hydrogen seen in its spectrum, that this hydrogen fuel was available in astonishing abundance despite its profligate consumption. A puzzle, certainly, was the origin of all this hydrogen, the fuel supply not only for our monstrously large Sun, but also for the vast

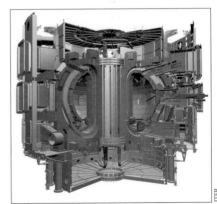

ITER concept, Cadarache

sea of stars making up our Galaxy, and the uncountable oceans of galaxies making up the Universe.

AN ANSWER, OF SORTS, came with the developments of Big Bang cosmology. Big Bang theory started life as a somewhat qualitative, hand-waving concept put forward to explain the origin of the Universe. Detailed calculations supporting its plausibility were, however, soon to follow. Fred Hoyle coined the moniker, during a 1949 radio broadcast, as a pejorative reference to a theory he did not believe.

Kolin Smith, New York

Ralph Alpher (1999)

Yet in a most remarkable application of scientific insight, detailed calculations of the interactions of subatomic particles within a few minutes of the Universe coming into being showed that conditions were just right for a few light elements to form as the temperature and density plummeted. This process of early element creation is referred to as primordial nucleosynthesis. The results of the first calculations were announced in 1948 by Ralph Alpher, Hans Bethe, and George Gamow.

These calculations tell us the following. Some three minutes after the Universe emerged from nothing, protons and neutrons were created. With them hydrogen, the simplest of all the elements, comprising just one electron and one proton, was formed in copious abundance. Cooked rapidly in the nascent gas were just a few other elements. The result of this primordial alchemy was about seventy five per cent hydrogen and twenty five per cent helium, trace amounts of hydrogen's heavier isotope deuterium, as well as a little lithium, and a smattering of unstable radioactive isotopes of tritium and beryllium. Just twenty minutes later, and the Universe was already too cool for any further similar reactions to occur. Nothing more could be concocted as the Universe struggled through its momentous birth pangs.

The predictions of Big Bang nucleosynthesis are strict, and by-and-large undisputed. The early Universe contained nothing else. No carbon and no oxygen, for example; no silicon nor phosphorous, no iron nor tin, no silver, gold, or uranium, just simply nothing of anything else. Observed abundances of light elements in the Universe are consistent with these predictions. The excellent agreement was the evidence needed to elevate the Big Bang hypothesis to the level of a theory. Later, it became evident that the density of normal matter predicted by Big Bang nucleosynthesis was much less than the mass of the Universe demanded by calculations of the expansion rate, but this inconsistency has been largely resolved by postulating the existence of dark matter.

SO IT WAS THAT THE UNIVERSE burst into existence with hydrogen enough to sustain successive generations of star formation stretching out over the sempiternity ahead. It was, however, only much later in the life of the expanding Universe, long after the Big Bang, that the enormous clouds of hydrogen gas, sinking into the gravitational wells of the mysterious dark matter, collapsed back to states dense enough for stars to form.

The next big advance in understanding was the realisation that the heavier elements seen throughout the Universe today result from other nuclear reactions within the stars. Only as the first stars started to burn according to the prescripts of nuclear fusion did the hydrogen within them slowly transform to helium. Later still, as the reservoirs of hydrogen inside a given star became exhausted, the helium started to 'burn' by similar fusion processes. Unlike the fleeting conditions of the early Universe, those inside the stars—the heavy element factories—remain hot and dense for millions or billions of years, time enough for

Fred Hoyle

slower nuclear capture processes to yield carbon, nitrogen and oxygen. Sir Fred Hoyle (1915–2001) was the pioneer of this field of stellar nucleosynthesis, while for his contributions and a seminal paper in 1957, William Fowler (1911–1995) was awarded the Nobel prize for physics in 1983.

Depending on their masses, some stars burn out and eventually collapse in on themselves and die, while more massive ones become ever more unstable until they explode in the unimaginable fury of a supernova. Born within the hot dense furnaces of the final cataclysmic stages of stellar evolution, the still more massive and exotic elements are created: in them are born all of the other elements: from neon to iron, cobalt to bismuth, polonium to uranium.

Let me stress again an important point: hydrogen and most of the helium in the world around us, including that making up all plants and animals, were created in the Big Bang fourteen billion years ago. All the other elements which make up the Earth, our bodies, and everything else, were cooked up and spewed out in the inner workings of long-dead stars. Everything else around us originated in these celestial furnaces. As Joni Mitchell penned in her 1969 song *Woodstock*, thereby reaching a somewhat larger audience than the more rigorous academic works of Alpher, Fowler and colleagues: "We are stardust, we are golden; we are billion year old carbon."

As the elements were created and hurled into the expanses of space from the explosive convulsions of the dying stars, they would be mixed into the unconsumed hydrogen and helium still residing in the interstellar gas clouds. As a result, star deaths lead to a gas composition no longer primordially fixed as just hydrogen and helium. Later stars are born from a new starting cosmic recipe, with a larger fraction of helium, a little more carbon, oxygen and silicon, and trappings of all other elements processed by the previous generation of stars from which they formed. The subsequent life cycle of stars which emerge out of this pre-processed gas are subtly but significantly different from those formed early on in the life of the Universe.

The branch of astronomy dealing with stellar structure and evolution is now immensely successful in modeling how a star with a given initial composition will evolve. Models predict how its internal structure will develop, and how long it will burn before its own death throes overtake it, and the cycle starts once again. Slowly, the hydrogen and helium in the Universe are being used up, but these cycles of star formation still have a very long way to run before all of the Universe's primordial fuel is consumed. Only many billions of years hence will the cosmic furnaces turn finally to an inert ash.

WE TURN THE CLOCK BACK to around 1910, and to another important discovery which played a key role in understanding the stars.

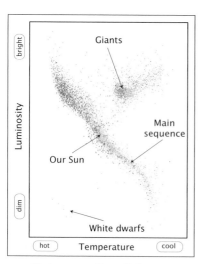

Hertzsprung–Russell diagram

Danish astronomer Ejnar Hertzsprung and American astronomer Henry Norris Russell discovered that a diagram of luminosity and temperature revealed a surprising relationship between the two quantities. This Hertzsprung–Russell diagram illustrates arguably the most important relation in all of astronomy. Far from being scattered randomly across the diagram, stars line up in swathes across it, deigning to occupy only certain well demarcated regions. The most dominant grouping, ranging from the hottest and most luminous stars to the coolest and dimmest, is called the main sequence. It is populated by all the stars which are in the longest phase of their lifetime—that in which the primordial hydrogen gas is being processed into helium. Our own Sun sits squarely within this sequence, very much an average sort of star. Above it, brighter and cooler stars are the giants in which helium burning dominates. Below it, a strip is populated by the dim white dwarfs, the final evolutionary state of the vast majority of stars whose mass is not too high. Elsewhere in the diagram, giants and supergiants, subdwarfs and brown dwarfs, all congregate in their favoured locations.

THE POINT SOMEWHERE along the main sequence where a star comes into existence is dictated by the mass of hydrogen gas which collapses to form it. To a much lesser extent it is also influenced by the small amount of elements more massive than helium contained within the same gaseous cloud as a result of previous generations of star formation. Thereafter, the lifetime of the star on this main sequence depends sensitively on its initial mass[45].

Perversely though it may seem at first glance, the higher a star's initial mass, the shorter its lifetime. This is simply because the heavier stars burn hotter and consume their nuclear fuel faster. Massive stars, ten times or so more massive than the Sun, survive for just a few tens of millions of years. The most massive, weighing in at a hundred times that of the Sun, will burn through and expire in three or four million years, the blink of an eye on a cosmic timepiece. Stars much less massive, around a tenth of our Sun's— like the ancient subdwarfs—can smoulder away for tens of billions of years, much longer than the present age of the Universe.

Our Sun itself, born nearly five billion years ago, will exist in its steadily burning state for another five billion years before its hydrogen fuel is exhausted, converted into helium. Fortunately for us, its luminosity and temperature are all but changeless over this immense gulf of time. But once the hydrogen fuel is exhausted, the next stages of a star's evolution proceed more rapidly as the heavier elements are fused in concentric zones within it. The consequences of its disappearing hydrogen will also be reflected in its changing position in the Hertzsprung–Russell diagram, changes which will inexorably gather pace. Some ten billion years after its formation, some five billion years hence, the Sun will move slowly to a point above the main sequence, pausing for a billion years or more while it burns through its gigantic helium reservoir.

Then things start happening ever more rapidly. At an age of twelve billion years its luminosity will increase rapidly, reaching a thousand times its present output, even though it will cool in the process from its present 5500 °C to around 3000 °C.

The next tumultuous stage lasts for only an instant of astronomical time, as the Sun, racing towards its last gasps, hurls off its outer gaseous envelope. Pulsations start, and quickly become more and more unstable. More of the atmosphere is ejected, progressively exposing deeper and hotter regions of its inner core, where temperatures top 200 000 °C[46]. The intense radiation

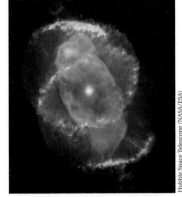

Cat's Eye planetary nebula

field then lights up the ejected tenuous atmosphere like a vast fluorescent tube. This phase is referred to, for historical reasons and somewhat misleadingly, as a planetary nebula.

All its fuel at last consumed, the Sun will have no outflow of energy left to support it, and it will surrender to the force of gravity, collapsing into a final white dwarf stage. The Sun will thereafter be nothing more than a hot cinder which slowly cools and dims, its once raging nuclear fires finally extinguished. At an age of thirteen billion years, eight billion years hence, the temperature will have fallen back to 8000 °C, and its luminosity will have dropped to a thousand times less than today. It will cool ever more slowly, its movement across the Hertzsprung–Russell diagram grinding almost to a halt. At fourteen billion years the temperature will be around 4000 °C. For billions of years thereafter, our all-but-dead Sun will just continue to glow like a stubborn ember.

THESE STAGES TRACED OUT for the Sun are repeated in broad outline for all stars. The details depends on the initial mass, to a lesser degree on the chemical composition, and to a still smaller extent on other properties: its rotation speed and its magnetic field being next in order of importance. The detailed evolutionary possibilities are many, akin to the complexities of biological classification branching its way from kingdoms and phyla down to the genus and species.

The hundred thousand stars mapped by Hipparcos provide a treasure chest in terms of its stellar census. Knowing the distance to a star is like being able to reach out and examine it in a terrestrial laboratory. Wide varieties of star types, with a wide range of masses, temperatures, chemical compositions and rotation rates, have been measured. We have, for the first time, accurate luminosities in large numbers. From increasingly sophisticated models, we have estimates of their masses and ages, and exactly where they are on their evolutionary paths. Counting their numbers, assessing their frequencies, and matching their distributions to models of the Galaxy, have provided a much finer picture of their inner workings, and the way in which they must have been assembled.

The detailed computer model predictions only match up with the star census if the physics of their interiors has been correctly described. Over the last decade, the models—much like those built to comprehend the Earth—have introduced more elaborate effects deep within. Included amongst them are turbulent flows driven by convection, stratification of various element layers, sedimentation of heavier elements towards the centre, and changing density profiles. Huge computers around the world run for days and weeks, mimicking their reactions and predicting their futures.

The final stages of a star's life are marked by all sorts of bizarre chemical signatures as it burns through and leaves successive layers of heavy elements in onion-like shells. Peculiar things also arise from competing effects of sedimentation pulling some elements to the centre, and radiation driving others outward. We can see, from their spectra, stars dominated by carbon or oxygen, others rich in barium and technetium, stars showing high levels of iron, chromium, manganese and nickel, and others distinguished by even more exotic elements such as scandium and yttrium.

A new technique, still in its infancy, allows us to probe the internal structure of the stars even more directly. The equivalent of seismology on Earth, it was first applied to our own Sun. By carefully following tiny brightness variations in its output over minutes and days, the Sun was seen to be vibrating due to vast energy turbulences deep inside—continuously ringing like an enormous bell when struck. Studying the very small light variations, changes of a few parts in a million, allow density variations deep within it to be charted.

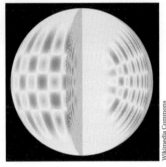

Seismology of a star interior

Applied to other stars, this new field of asteroseismology is starting to tell us more directly about their internal structure—how their density changes with depth, and how they rotate far within. The predicted behaviour depends strongly on the star's age and evolutionary state. From the distances to these carefully-studied stars we now have detailed estimates of their energy output. These can be compared with their own particular vibration frequencies to see whether the full model of the star's internal structure and evolution hang together. So far, they do.

Although these details are not important for this story, what is important to convey is the very wide range of processes going on inside the stars. Many different characteristics and dependencies result. In so many remarkable ways, the vast diversity of stars that we can see and survey are now very well replicated by our theories.

FROM THE COMPUTER MODELS, and great insight too, we find explanations of why some stars have a very well determined energy output over parts of their lives, or conversely why some are highly variable, pulsating hugely in intensity over days or weeks. This has helped to identify astronomy's 'standard candles', objects whose average brightness seems to be fixed—and explicable—by the laws of physics. We can then get an estimate of their distance from their observed brightness combined with their actual luminosity.

NASA/KAO/Adrian Pingstone

The Large Magellanic Cloud

Important amongst them are the highly-variable stars called Cepheids, bright young stars formed in relatively recent bursts of star formation in nearby spiral arms, and the RR Lyrae stars, comparably bright but very much older. These variable stars are hugely luminous, some of the brightest in all of creation. They are so distinctive that they can be seen shining out and pulsating regularly like lighthouses far away across the distant expanses of our Galaxy, even amongst the millions of others stars in the same direction. Since their discovery nearly a hundred years ago, they have provided the best celestial rulers to plumb the vast distances across our Galaxy. They give us a distance to the centre of our Galaxy, and the key to estimating the gulfs of space to the galaxies next beyond our own, the Large and Small Magellanic Clouds, and beyond.

In this way, this modest thirty centimeter diameter telescope has given the most secure distance to the Large Magellanic Cloud to date. Estimates pin it down to a little short of fifty thousand parsecs, some two hundred thousand light-years. The diffuse sources of light that we can see with the unaided eye from the southern hemisphere, actually left the Large Magellanic Cloud during the geological period of the Middle Pleistocene on Earth. These rays began their journey just as modern humans were making their first appearance in fossil records in Africa, yet to start out on their worldwide migration.

WE MET THE WHITE DWARF STARS when describing the final fate of our Sun, still billions of years hence. A white dwarf is just the inner core of a once-raging star, now inactive, all nuclear fusion nothing more than history. Our theories of star evolution tell us that nine out of ten stars will end their lives in this way. Only the most massive will leave the cosmic stage, more spectacularly, as supernovae.

White dwarfs are dim, and difficult to find amongst the huge numbers of faint and more distant stars. Some stand out because of their proximity and high angular motion on the sky, such as Van Maanen's star, the closest known solitary white dwarf discovered in 1917 by Adriaan van Maanen. The term white dwarf was coined in 1922 by Willem Luyten, the first PhD student of Ejnar Hertzsprung in Leiden, and later renowned as a highly successful hunter of many other high proper motion stars.

White dwarfs figure large in astronomical mystique because of the exotic nature of their matter. Fusion having ceased, the nuclear heat for so long generated inside the star can no longer support it, and no longer prevent it from its inexorable gravitational collapse. The dead star's mass, still comparable to that of the Sun, falls in on itself until it reaches a small dense sphere the size of the Earth. The resulting density is a million times that of rocks on Earth. A cubic meter of white dwarf carbon would weigh in at a hefty million tonnes, an astonishing one tonne to the cubic centimeter.

Willem Luyten

What stops it collapsing further is an effect labeled by those in the know as electron degeneracy pressure. It's a consequence of the Pauli exclusion principle of quantum mechanics, which holds that no two electrons can occupy the same quantum state. This incomprehensible jargon should not perturb the reader too much. A white dwarf is simply held up by a special kind of force not evident in our daily toil, one that sets a limit on how much matter can be squeezed together once gravity dominates. Just as the Earth doesn't collapse under its own weight due to other (electromagnetic) forces—which we happily accept because the result parades before us—so these huge gravitational forces just compress matter beyond anything normally occurring around us, causing it to be extremely dense. If the mass of the dead star exceeds about one and a half times the mass of the Sun, incidentally, even this exotic pressure can't hold it up—the matter is squeezed into a terminal collapse, swallowing itself in a self-generated black hole.

Ten years devoted to observing peculiarities in the motion of the bright star Sirius led Friedrich Bessel, in 1844, to conclude that it had a dim companion, later shown to be a white dwarf. Meticulous computation of its six-year orbit around Sirius by Christian Peters in 1851 led to the prediction of its position, and its discovery in 1862 by Alvan Clark. Its density was estimated some fifty years later, giving a number so high that Ernst Öpik, Estonian grandfather of British MP Lembit Öpik, who calculated it considered it to be simply implausible.

As Arthur Eddington put it in 1927 "We learn about the stars by receiving and interpreting the messages which their light brings to us. The message of the companion of Sirius when it was decoded ran: 'I am composed of material three thousand times denser than anything you have ever come across; a ton of my material would be a little nugget that you could put in a matchbox.' What reply can one make to such a message? The reply which most of us made in 1914 was 'Shut up. Don't talk nonsense.' "

Sirius is the brightest star in the sky, and there are even reports of it having been observed in daylight by eye, although only under exacting conditions. It lies south along the line of the three bright stars forming the belt of Orion, as far to the southeast as the Hyades cluster is to the northwest. Stare up at it by all means, and try the impossible of imagining the white dwarf in orbit around it.

University of Chicago

Subrahmanyan Chandrasekhar

The theory of matter at these tremendous densities was worked out by astrophysicist Subrahmanyan Chandrasekhar (1910–1995) in 1931, and refined by others over the following half century. For his work on the structure of stars, Chandrasekhar was awarded the Nobel prize for physics in 1983, jointly with William Fowler.

His theory predicted that the mass of a white dwarf would be related to its radius in a very specific way. Yet surprisingly for such a radical theory, which forms a central assumption underlying a whole range of studies of white dwarf properties, this relationship has long remained largely untested by observation—too few white dwarfs have well-determined masses and precisely-determined radii. This is despite the fact that so many stars have exited via the white dwarf route, the tell-tale legacy of the fiery suns that once burned within them, that the number of these dying embers populating the stellar graveyard is enormous. More than ten thousand are now known.

Yet they are amongst the very dimmest of the stars, and not at all easy to detect in the numbers representative of their true ubiquity. Hipparcos was able to observe just twenty two: half of which are solitary travelers through space, the other half are in orbit around other stars. Though a sample only very small in number, Hipparcos delivered far better measurements. When the first results of its white dwarfs studies hit the scientific press in 1997, they were received with considerable excitement.

For some of the objects, the satellite's exquisite distance determinations perfectly validated the theoretical predictions. For the white dwarf in its fifty year orbit around Sirius, for example, its mass turned out to be almost identical to that of our Sun, while its radius is a little less than that of the Earth, at just 5800 kilometers. It fits Chandrasekhar's theory perfectly if it has a carbon core covered by a thin dense sea of hydrogen.

However, Judith Provencal, from the University of Delaware, showed that three of the Hipparcos white dwarfs, including one around Procyon, gave sizes too small for their measured masses. Procyon is one of the nearest stars to Earth, at a distance of just over three parsecs or about eleven light-years. Its white dwarf orbits with a period of forty years, and at a separation roughly that between Uranus and the Sun. Its existence, though not its white dwarf nature, was known from the star's wobbling orbital motion already in 1861. Although somewhat larger than the white dwarf around Sirius, it's

Judith Provencal (2007)

a lot cooler. It's possible that these objects have iron rather than carbon cores, but extremely compact 'strange dwarfs' comprising quark matter at even higher densities have also been proposed. The evidence has shifted back in favour of a carbon core for the white dwarf around Procyon, while 'strange' cores remain a possibility still being debated for the others.

Most astronomers would probably put their wager on these strange white dwarfs being nothing more than red herrings, but it's a possibility that Hipparcos has made the first discovery of a superdense quark body. In this sense, we have a snapshot of science in action: a novel idea which might simply fade after a false dawn once better data are acquired, or one which might blossom into a vibrant new research field—so obvious, one way or the other decades hence, to everyone blessed with great hindsight.

THAT WHITE DWARFS ARE DEAD WORLDS of hot carbon left hanging in space, visible only through their residual heat stored within them, provides us with a most unusual type of clock. Their size and huge mass means that their cool-down, as they radiate their heat into the freezing depths of space, takes billions of years. So we can get an age from their temperature, in the same way that a pathologist can estimate the time of death from the temperature of a corpse. In practice, white dwarfs are found spanning a vast range of temperatures, from the 200 000 °C of the recently deceased, to the 4000 °C of the most ancient stellar corpses which expired billions of years ago.

This incredibly slow cooling neatly explains why no white dwarfs have been found with temperatures below 4000 °C. If the Universe is truly less than fourteen billion years old, timed since the purported Big Bang, no stars within it can possibly be older. Naturally, no white dwarfs, which formed from dying stars, can be older than their own progenitors either. The temperatures of the coolest white dwarfs ever found hang nicely together with the age of the Big Bang. None have had time to cool further, and nothing that we are aware of predates this moment in time.

Comte de Buffon

If we are on the wrong track with Big Bang cosmology, it would be a strange coincidence indeed if lower temperature white dwarfs are not out there, lurking somewhere, and a doubly curious conspiracy if they exist but none have so far been found.

This method of age-dating has a distinguished pedigree, since similar methods were once used to estimate the age of the Earth[47]. Isaac Newton (1642–1727), German polymath Gottfried Wilhelm Leibniz (1646–1716), French naturalist Georges–Louis Leclerc (Comte de Buffon, 1707–1788), and Ulster Scot physicist William Thomson (Lord Kelvin, 1824–1907) all took turns in deriving the age of the Earth from its estimated cooling time. Buffon, for example, is acclaimed for his enyclopaedic *Histoire Naturelle, Générale et Particulière* in 44 volumes, in which he set out to synthesise all available knowledge of nature and natural history. In *Les Époques de la Nature* in 1778 he described how he derived an age for the Earth by heating ten iron spheres of different sizes until white hot, and then observing the time taken for them to cool. Extrapolated to the size of the Earth, which he conjectured had started out as a molten sphere, he arrived at an age of just 75 000 years.

SO FAR, WE HAVE LOOKED only at single stars. Although many stars seem to have been born alone, albeit often in clusters with siblings formed from the same interstellar gas clouds, many are born as twins (or binaries), triplets or even quadruplets. The tenuous swirling gas may collapse not into a single nuclear furnace, but into two or more stars orbiting around each other, like partners dancing around a ballroom. In much the same way, stars in a binary pair swing equally from side-to-side if their weights are equal, while a much heavier companion is only pulled aside a touch if the weights are more unbalanced. Our Sun is not immune, and itself swings around a little as the planets of the solar system tug at it in their various orbits.

Since the invention of the telescope, many close double stars have been found. Mizar, in the constellation of Ursa Major, was observed to be double by Giovanni Riccioli already in 1650, and possibly before. It's the second star from the end of the Big Dipper's handle, and the ability to discern nearby Alcor with the naked eye is often quoted as a test of good eyesight. John Michell (1724–1793) suggested that some double stars might be physically associated, rather than just chance alignments, one perhaps far behind the other. William Herschel began observing double stars systematically in 1779, and duly published catalogues of some seven hundred. He observed changes in the relative positions in a number of double stars over the course of twenty five years, and concluded that they must be true 'associated' binary systems. The first orbit—a precise mathematical description of how each component moves microscopically back-and-forth on the sky—was computed in 1827 by Félix Savary.

Today, more than a hundred thousand binary stars have been catalogued, and orbits are known for thousands. Within our Galaxy there are billions. They exist in a bewildering plethora of design. They range from very close pairs in which the stars are all but in contact, separated by a few times the Sun's diameter and orbiting each other in a frenzied interval of just a few hours—one can only imagine living on another Earth in which such a chase appeared daily in front of our eyes. Some pairs are so close, and so unbalanced in mass and size, that gas is continuously being stripped from one by the gravity field of the other, to fall to its surface with the force of endless nuclear explosions, hurling out streams of intense X-rays as the material smashes down into it.

Mizar and fainter Alcor

Digitized Sky Survey

At the other extreme are pairs which remain only precariously attached by this invisible thread, the rapidly diminishing force of gravity. They can be separated by more than ten thousand times the Earth–Sun distance, one tenth of that to the nearest star. Their orbital periods can exceed a hundred thousand years.

The importance of binary stars throughout astronomy is surprisingly profound. They provide important checks on theories of star formation, and crucial handles on stellar evolution. But towering above all else is the unique opportunity that they provide to measure a star's mass. For a single star, hanging alone in the nothingness of space, there is simply no possibility of weighing it. We cannot determine its mass from its luminosity, from its age, or from its spectrum, and yet its mass is one of its most fundamental of properties, determining how it burns its fuel, and how it evolves with time. Only when one star is trapped by another in a binary system can we measure the orbit, appeal to the role of gravitational attraction between them, and get a measure of their masses. We weigh the Sun in just the same way: based upon the one year period of the Earth's orbit, and the distance from the Earth to the Sun.

Accurate star masses are highly prized, and only a few dozen are known more accurately than a few per cent. An application of Kepler's Laws of motion shows why this is so: while star distances are hard enough to measure well, star masses in a binary pair depend on the third power of the distance. As a result, unreliable distances to the stars quickly translate into very unreliable masses. Even quite good distances translate into not very good masses. Hipparcos observed many thousands of known binaries, improving their distances, orbits, and masses, and improving our knowledge of their ages and their internal structures. And careful mass measurements for a few then give useful approximations for other stars of similar type.

IT HAD BEEN EXPECTED that the Hipparcos observations would reveal many previously unknown binary pairs, and this indeed turned out to be the case. At the heart of the measurement principles, the repeated satellite observations picked out not only the motion of the stars through space, but also their wobbling parallax motion as the Earth orbits the Sun. Binary stars add yet another type of wobble on top, due to the fact that the two stars are circling each other. If the conditions were favourable, and the orbit period not too long, this extra wobble could be detected, and new binaries came tumbling out of the computational number-crunching in their thousands.

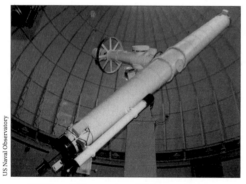
US Naval Observatory

Yet more were found in copious numbers once the catalogue was published. New binaries came from observations of selected Hipparcos stars using special telescopes on the ground designed to zoom in on them at the highest angular resolutions—in these special observations, American and Russian astronomers have led the way.

The US Naval Observatory serves as the official standard of time for the entire United States. Established in 1830 as the Depot of Charts and In-

US Naval Observatory 26-inch refractor

struments, it was made into a national observatory in 1842. Mapping the stars has always been central to its mandate, and binaries have been observed using its twenty-six inch refractor in Washington, with few interruptions, since 1875. Once the largest telescopes in the world, it was used by Asaph Hall in his discovery of the satellites of Mars, Phobos and Deimos, in 1877. Thousands of Hipparcos binary stars have been observed over the last few years by this historic instrument.

On the other side of the world, Russia's Pulkovo Observatory, outside Saint Petersburg, was set up around the same time, in 1839, dedicated to timekeeping, weather, and navigation. In the 1950s the Soviet Academy of Sciences set the goal of building the world's largest optical telescope, a title long held by the five meter Hale telescope at Palomar Observatory in California. Designed out of Pulkovo, the six meter telescope of the Special Astrophysical Observatory started observing in 1975. Now equipped with modern CCD detectors and a special high-resolution camera, Yuri Balega has led the new measurements of many Hipparcos binaries from this two thousand meter high site, deep in the northern Caucasus mountains.

Yuri Balega (2007)

More binaries were thrown up after adding ground measures of their radial motions. Still more came by analysing the details of stars that showed a discrepancy between the motions measured by Hipparcos, with those measured over decades from the ground—many of these were found to be orbiting systems with long periods of fifty years or more. And still others came by examining the motions of stars observed for the best part of a hundred years from the ground as part of the long-term monitoring of the Earth's rotation.

HIPPARCOS CONFIRMS THAT BINARY stars are extremely common, and perhaps more than half of all stars occur in binary or multiple systems. Because it's so difficult to detect binaries with long periods of centuries or more, or those with very dim companions swamped by the light from the dominant star, it's possible that essentially all stars may have a companion—perhaps, even, also our Sun.

Back in 1984 came a hint that our Sun has a companion star in orbit around it. Paleontologists David Raup and Jack Sepkoski, at the University of Chicago, identified a periodicity in extinction rates on Earth over its geological history. From the extinction of fossil families of marine life and protozoans, they identified twelve extinction events over the last two hundred and fifty million years. The average interval between extinctions was twenty six million years, and they speculated about a non-terrestrial cause.

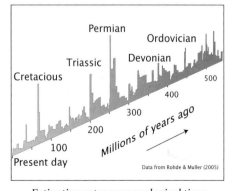

Extinction rates over geological time

The challenge to find a mechanism was picked up by two teams of astronomers, who quickly and independently put forward the hypothesis that the Sun may have an as yet undetected companion star. If the companion is in a very elliptical orbit, it could disturb comets in the Oort cloud each time it comes close to the Sun, causing a periodic increase in the number of comets visiting the inner solar system, with a consequential increase in impact events on Earth. The idea became known as the Nemesis, or Death Star, hypothesis. A very different possibility is that any periodicity is caused by the Sun oscillating across the plane of the Milky Way.

That the Sun may have a companion star might sound unlikely, but the Hipparcos census cannot rule it out—it could be a dim companion, in a very wide orbit, and with an extremely long period. If it exists, which is far from certain, then it must be faint, and identifying it will not be easy. But history points to some precedents. It might already be reposing in an existing star catalogue masquerading, like Groombridge 1830 did for a quarter of a millennium, as an object of no particular scientific interest. But it would be unmasked if its distance could be measured. Hipparcos did not pick it out, so it was not in the initial catalogue of this first space census. But perhaps, even now, it's racing through space on course for our solar system, and another encounter with us just a few million years hence. The next leap in space astrometry will spot it if it's there.

ASSOCIATIONS ARE COLLECTIONS of young stars only recently born from a localised burst of star formation somewhere within the Galaxy's spiral arms. They are Nature's stellar nurseries, and remarkable laboratories for studying the very earliest stages of star formation. The Hipparcos survey included several thousand stars recently born in nearby star-forming regions, all within a few hundred light-years of our solar system.

These stellar nurseries are the very places in which binary systems are formed. But the star densities can be high enough that the reverse may also happen: a binary star may pass very close to another binary star, and a remarkable manifestation of the complexities of gravitational interactions then enacted: one pair is forced closer together, while the other pair is torn apart. Through a kind of sling-shot mechanism, the two disrupted stars shoot off at high velocities, in two very different directions. Such strange objects are called runaway stars. Surveys of swiftly moving stars in the 1950s and 1960s already discovered a number. Those that were known were carefully included in the Hipparcos programme so that a much more detailed scrutiny could be made once the catalogue was published.

A fine example are two massive young stars with the magisterial names of AE Aurigae and μ Columbae, which lie some seventy degrees apart, on almost opposite sides of the sky. They are so well separated that one would confidently predict that they must be unrelated. But a careful inspection shows something very different, for they are moving away from each other with almost equal and opposite speeds of around a hundred kilometers per second, nearly half a million kilometers an hour. It is as if they had emerged from the same point in space, millions of years ago. Their common origin had already been suspected by Adriaan Blaauw in 1954, and Hipparcos gave a new possibility to examine their provenance.

Ronnie Hoogerwerf

Ronnie Hoogerwerf, a PhD student working in Leiden, traced back their motions through time and space, and showed that both stars were ejected at the same time from the Trapezium star cluster in the constellation of Orion, a little more than two million years ago. At their birthplace is a binary known as ι Orionis, which seems to have been left behind as the survivor of such a very close fly-by of two binary systems. AE Aurigae and μ Columbae were the two ejected stars of the other disrupted binary. Ever since, they have been hurtling through the Galaxy, a pair of Nature's most awesome cannon balls, racing across the sky to their present-day positions. The former crossed the constellation Taurus some million years ago, thundering northward on its way into the constellation of Auriga. The latter set out through Lepus, haring its way south into the constellation of Columba. Many new ideas about star formation and the disruption of binaries have come from these long-separated twins. A number of other runaway stars have also been found hiding in the catalogue.

AE Aurigae and μ Columbae are traveling at super-high velocities, but even they will not have enough energy to escape from the gravitational clutches of our Galaxy. Their travels have already taken them more than 10 000 000 000 000 000 kilometers, and they will go much further still over the millions

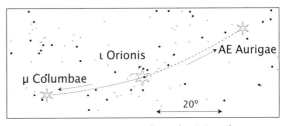

Escaping runaway stars from the Orion cluster

of years to come. Yet their struggle to escape the Galaxy will have been pointless, and the outcome of their vain attempt can be predicted with confidence—they will gradually slow to a halt, unwillingly turn around, and from some spectacular vantage point high above the Galactic plane, set off again on their lonely journey home. With no additional impulse, an object on Earth demands an escape speed of about eleven kilometers per second to break free from our own gravity field. In the same way, an object in our Galaxy—whatever its mass—needs a speed of about five hundred kilometers per second before it can escape into intergalactic space. Few stars are so well endowed, and few will therefore ever escape. Only ten stars out of the entire Hipparcos catalogue have speeds above three hundred and fifty kilometers per second, and none exceed the Galactic breakaway speed limit.

The distribution of star speeds in our neck of the woods actually provides a handle on the total mass of our Galaxy. Doing the sums, our entire Galaxy seems to weigh in at the equivalent of about five hundred billion times the mass of our Sun. Again, it's surprising and ingenious, and satisfying for those that made the survey, that by measuring the positions of the stars around us we can weigh the Galaxy. It's also a good moment to recall, lest we forget, that all of this mass—as well as all that in all the other galaxies—appeared out of nothing thirteen billion years ago.

ONE OF THE THINGS that makes big-game science so exciting is that, just when you think you've discovered and captured some of the Universe's strangest and most bizarre animals, like the runaway stars, along comes something even more extreme and unexpected. So it was that in 1988 in the journal *Nature*, Jack Hills, at the Los Alamos National Laboratory in New Mexico, published a prediction that a tightly-bound binary star approaching close to the vicinity of a black hole would be disrupted, and its components hurled apart with even greater fury than in the binary–binary near collision just described. The prediction was important because of the mounting evidence that there lies, at the very centre of our own Galaxy, just such a black hole. Current measurements suggest that this monstrosity weighs in at more than a million times the mass of our Sun.

This may sound alarming, it may read like science fiction, and it may appear to be completely hypothetical. But a similar disruption process playing out in the centre of the Galaxy could then leave one component of the binary trapped in orbit around the black hole, and the other ejected at extreme velocities of up to four thousand kilometers per second. The discovery of even one such high-speed star coming from the Galactic centre would provide, so the study argued, nearly definitive evidence for a massive black hole. It's a fine example of an object never having been seen or suspected, suddenly being predicted based on other knowledge.

The first of these so-called 'hypervelocity stars' was duly discovered from the bountiful archives of one of the grandest surveys of the heavens carried out from ground, the Sloan Digital Sky Survey. Starting its own patrol from a special telescope at the Apache Point observatory in New Mexico in 2000, it is surveying a hundred million stars to provide their positions, colours, and spectra. In 2005, a faint and otherwise quite unremarkable star, with no discernible motion across the sky, was found to be the harbinger of astonishing tidings. Its spectrum, encoding its Doppler velocity shift, announced that it was traveling away from us with an unprecedented velocity of a little less than a thousand kilometers a second, nearly three million kilometers an hour.

Tracing its known flight path backwards in time and space, we can say that it was ejected from the centre of our Galaxy around eighty million years ago, just as dinosaurs were disappearing on Earth. It now lies nearly 200 000 light-years distant, far beyond the limits of our Galaxy. It is still racing on outwards, its own struggle against our Galaxy's tidal forces already won. A number of other such stars are now also known from the same survey. Although these objects are too faint to have been part of the Hipparcos catalogue, they are examples of new, bizarre types of star only recently discovered from their space motions, fodder for the giant deep-sky high-accuracy astrometric surveys now being planned for the future.

IF THE STAR DENSITY in these young stellar nurseries is sparse, the stars born in them will spread out, and dissolve quickly back into the surroundings, becoming pretty much indistinguishable from other field stars after just a few million years. But if enough are born at once, leaving them bound together by their overall gravity, the resulting open clusters may persist as recognisable concentrations of a few hundred stars for tens or even hundreds of millions of years. Open clusters, I should stress, are very different from the globular clusters, the ancient and more dense remnants of captured galaxies torn apart as they fell into our Galaxy long ago.

More than fifteen hundred open clusters are known. The fact that they often lie buried within the dense star factories of the spiral arms means that there are many more still to be found. If our local region of the Galaxy is representative, there could be a hundred thousand other open clusters ranged throughout its vast entirety. As stellar evolution and gravity disruptions take their toll, these clusters also slowly disintegrate and disappear. As the spiral density waves trundle on round, more are born.

Just two of these open star clusters are visible to the naked eye, and only then with a little care. To find them, the winter sky in the northern hemisphere, dominated by the constellation of Orion, is the place to start. Follow the three bright stars of the constellation's 'belt' northwards for twenty degrees or so—a full span of a hand at arm's length—to reach the bright orange-tinged star Aldebaran. There, the tight V-shaped grouping of stars are some of the brightest members of the Hyades cluster. Aldebaran, alone amongst the bright stars in this region of the sky, is not a cluster member, but sits in the foreground at a distance of about sixty light-years. A little further northwest, along the same line, lies the conspicuous and very compact group of blue–white stars known as the Pleiades, or the Seven Sisters. Both clusters lie in the constellation of Taurus, straddling one either side of the ecliptic, that imaginary line across the heavens along which the Sun and planets appear to move.

The Hyades cluster from the Hipparcos catalogue

If you try to find these clusters in the night sky, compare them with the computer-generated images of the regions given on pages 268–269. Try to imagine them, not as stars projected flat on the sky, but with some true distribution in distance, spread out a little along the line-of-sight. The stereo images reconstruct their true space distributions, a knowledge of which has been essential for the investigations which follow. The Hyades is the closest of the open clusters at about a hundred and fifty light-years distance. It's a collection of a few hundred stars packed into a volume about thirty light-years in size. The relative proximity of the Hyades means that its members appear spread over a considerable area of the sky—some ten degrees or more. Obvious members near the central regions merge progressively into the hazy background of unrelated stars further out. In the case of the more distant and more compact Pleiades cluster, half a dozen are visible to the naked eye, while many more appear through small binoculars. Both clusters are fine sights on a still dark night, and must have been arresting sights in the dark skies of antiquity. According to some recent authorities, the famous painted caves at Lascaux in central France, decorated by the hand of prehistoric man some 17 000 years ago, include a depiction of the Pleiades cluster.

Unlike the constellations which are coincidental patterns of stars as seen from Earth, star clusters are real physical entities. Their member stars are held together by that recurring master of the Universe, the pull of gravity. Because the stars are gravitationally bound, they move through space together, and this means that they also move across the sky together, albeit by a rather tiny amount each year. If we could attach an arrow to each cluster member to show its direction and speed of motion, arrows of the members would all point the same way and all would have the same speed. The non-members, whether foreground or background stars, would have arrows of different lengths pointing in different directions, a manifestation of their random motions through space.

Before Hipparcos, a first stab at determining membership could be made from their space motions measured from the ground over many decades. This in turn led to a way of estimating the cluster's distance based upon its perspective motion, a clever trick first applied to the Hyades cluster by American astronomer Lewis Boss in 1908. A reasonable mean distance could be estimated for each typical cluster member, and these could be used as a kind of standard candle

The Pleiades star cluster

for other clusters of stars found much further away. For a hundred years before Hipparcos, the Hyades cluster was an important step on the rickety ladder used to scale the distances to stars far beyond. But it was not always certain which of the fainter stars were members and which were not, and this was a big complication.

HIPPARCOS OBSERVED A COUPLE OF HUNDRED of the four hundred or so stars believed to be members of the Hyades cluster. It measured their motions across the sky, and their distances directly from their parallax. The two measurements taken together provide a powerful way of defining cluster membership, and of figuring out finer details of their motions. We could refine the membership list, and rule on which stars were members of the Hyades club, and which were not. From the distances we could reconstruct the three-dimensional structure of the entire cluster in space. From the space velocities we can see how the stars are rattling around within the cluster. Just as the components of a binary star turn about their common (but invisible) centre of mass, and just as the stars within our Galaxy disk oscillate up and down about the (invisible) centre of mass defined by all of the stars concentrated around the plane, so too do all stars in a gravitationally bound cluster oscillate back and forth around their centre of mass defined by the hundreds of stars within it.

An open cluster is far from reaching the very high star densities found in globular clusters. If we found ourselves transported to a planet circling a star within the Hyades cluster, dancing in the gravitational well of the other members, the night sky would be a little more resplendent than our own—three or four hundred more stars would be arranged around us, their brightnesses vying with the brightest in our own night sky. So finely balanced is the gravitational poise of all the stars in the cluster, that they would be moving slowly one with respect to the other with a relative speed of barely a thousand kilometers an hour, coasting through their assigned volume of space at the speed of a commercial aircraft. Each would be exercising its own dance back and forth through space, taking a million years or so to stroll across to the other side of the cluster, before being pulled back again by all the other members of the celestial ballet, all over again to continue interminably its own tireless adagio.

The Hyades cluster was born hundreds of millions of years ago, and has been riding around confidently in the plane of Galaxy since then. It has its own collective motion through the Milky Way of about fifty kilometers per second. Does this grand collection of a few hundred stars hurtling together through space trample over everything in its passage? Do other stars in the disk move out of its way as it looms up out of the darkness and bears down on any star unfortunate enough to lie in its path?

Curiously, the answer is no. Space, as we have seen, is phenomenally empty, and even the most crowded open clusters are in reality only just a little less empty. A cluster glides through space, making its ghostly passage through the Galaxy without impact. Even two such clusters bearing down on each other head-on would simply pass straight through one another, a physical encounter between any of the stars sufficiently improbable to disregard.

The stars in such a cluster inhabit its self-made gravity well, and the stars by-and-large remain trapped inside it for a large fraction of eternity, at least for hundreds of millions of years. During this time, some will burn out, and some will collapse into white dwarfs. Yet a few stars may escape. At certain parts of its travels around the Galaxy disk, a cluster will occasionally come close to other massive objects, such as giant gas clouds. At these points the gravity field of the cluster might be closely counterbalanced by that of the nearby object, providing a sort of gravity tunnel for a few stars to stumble out through. Around the central regions of the Hyades cluster we can see a few of these lonely escapers, together forming a delicate halo of a few stars moving outwards from its deep gravitational well. As more escape, the gravitational pull of the others weakens, the break-up and ultimate demise of the cluster hastens, and the curtain eventually falls on its long and magnificent performance.

The most important clues of the inner workings of star clusters come from comparing the luminosity of the stars with their temperature, their Hertzsprung–Russell diagram. The Hyades stars line up perfectly in this arrangement, and our detailed stellar models can be tuned to replicate their behaviour. Since all the stars in any one cluster are assumed to have been born together, out of the same cloud with the same chemical composition, there is not much freedom to adjust the models. What drops out is that the Hyades cluster was born 625 million years ago, from a gas cloud with a heavy element composition a little higher than that of our own Sun.

The Pleiades cluster, nearby on the sky and the next nearest to us in distance, raises some more awkward questions. Being further away, at a little more than a hundred parsecs, around three hundred and fifty light-years, the distance measurements for each star aren't quite up to the quality of those in the Hyades. It's just on the edge of the distance horizon for which the Hipparcos parallaxes are of their best. We can still nail down a distance for the cluster, and we can fix its age to about a hundred million years. But its distance estimated from our best stellar models doesn't fit particularly well with the distance measurements made by Hipparcos. The discrepancy between these two nearest clusters doesn't yet allow us to figure out in what way our current models could be wrong.

BASED FIRMLY ON OUR NEW STAR POSITIONS, we have a set of ages given by stellar models, including ages for open star clusters, and ages for the very oldest stars that are known anywhere in the Universe. Their relics, the white dwarfs, litter our Galaxy and give another upper age limit from their enormous long cool-down times. We can assign ages to the oldest of the ancient globular clusters of around ten or twelve billion years. Beyond these, nothing has been found which is older.

Meanwhile the accurate distance now measured out to the Large Magellanic Cloud is being used, in turn, to calibrate the brightness of still more distant galaxies. These distances are used as the foundation of cosmology to set the scale of the expanding Universe. Taken together with the velocities of the receding galaxies far from our own, they too give a figure of about 13.4 billion years for the time since the Big Bang.

So it is that these various ways of dating the Universe come together, and leave us rather secure that we do know its age, and that we have taken the first faltering steps to comprehend its origin.

"Give me a lever long enough and a fulcrum on which to place it, and I shall move the world," Archimedes is quoted as saying. "Give us a mirror polished enough and a platform on which to place it, and we shall date the Universe," is our twenty-first century riposte!

Chapter 13

Our Solar System and Habitability

You teach your daughters the diameters of the planets and wonder when you are done that they do not delight in your company.

Samuel Johnson (1709–1784)

OUR SOLAR SYSTEM IS DOMINATED BY THE SUN, by far the most massive body within it, and around which all other solar system objects orbit. Far from being an isolated star ploughing its way alone through the vast expanses of the Galaxy, there are many smaller objects under its direct and eternal gravitational control which accompany it on its celestial journey.

There are the eight major planets, or nine for old-timers who still count Pluto amongst them. There are several dozen planetary satellites, including our own Moon. There are hundreds of thousands of known asteroids (or minor planets) a kilometer or more in size, along with countless millions of smaller objects, still uncatalogued, roughly occupying the region between Mars and Jupiter. Some interlopers, the near-Earth objects, have been dislodged from their circular orbits and venture much closer to Earth.

Impacts with Earth are common. Although most small bodies reaching its upper atmosphere disintegrate and burn up on their way to the surface, several hundred meteorites ranging in size from a few centimeters upwards are estimated to reach the surface each year.

M. Perryman, *The Making of History's Greatest Star Map*, Astronomers' Universe
DOI 10.1007/978-3-642-11602-5_14, © Springer-Verlag Berlin Heidelberg 2010

Wikipedia Public Domain

Comet McNaught, 2007

Beyond Neptune are the recently-found family of trans-Neptunian, or Kuiper belt, objects. Further out still lies a bleak twilight wilderness of icy objects forming the enormous shroud of the cometary Oort cloud. These are occasionally forcibly jostled into orbits bringing them much closer to the Sun. The increasing warmth heats and evaporates their icy mantle as they approach our own celestial furnace, resulting in the occasional spectacular cometary displays. Some plunge directly into the Sun, immediately engulfed by the overpowering enormity of its gravitational attraction. Others may circle the Sun in relatively stable short or long-period orbits, like the seventy six year orbit of the renowned Comet Halley, or the great comet of 2007, Comet McNaught.

OUR OWN SOLAR SYSTEM slowly emerged from the mists of time and space, a shade over four and a half billion years ago. From an enormous cooling cloud of molecular gas, still mostly hydrogen and itself a remnant of the Big Bang almost ten billion years before, structure slowly appeared. Under the pull of its own gravity, and because of its rotation, the collapsing gas cloud formed into a flattened disk, like a giant wheel turning slowly about its central axis. Far out from the Sun itself, circulating gas and dust slowly collided and progressively stuck into ever larger bodies, first into dusty flakes, and then into bigger clumps. Over millions of years these too collided and stuck together to form still larger planetesimals, tens of kilometers in size. Some of these larger planetesimals progressively formed the Earth, and the other rocky inner planets that accompany us in our orbit around the Sun.

The planets farther out, the gas giants—Jupiter, Saturn, Uranus and Neptune—probably also started life as small rocky cores. But as they moved round and round the Sun, they swept up any remaining gas, accumulating any left-over hydrogen from the disk lanes that they gouged out. This progressive accretion eventually resulted in small central rocky cores being surrounded by oceans of metallic hydrogen. In the case of Jupiter, its gaseous hydrogen sea is more than seventy thousand kilometers deep. The huge gravity fields of these colossal gas giants trapped satellites of their own, their own moons, as other less massive bodies ventured too close.

The asteroids, or minor planets, are remnants of this early protoplanetary phase. They are relatively small, mostly a few kilometers or less across. They occupy a region of the once colliding disk in which the early stages of accretion was slowed by the large gravitational force of Jupiter. The Moon, Mercury, and the Martian moon Phobos, amongst others, bear spectacular surface craters, testament to the impacts that dominated the early formation of the solar system billions of years ago. Our Moon formed around this time, probably as a result of a Mars-sized body hitting the proto-Earth, hurling material into orbit from which it duly solidified.

Phobos, moon of Mars

Many asteroids are tracked, and a knowledge of their orbit is maintained, by the Minor Planet Data Center at the Smithsonian Astrophysical Observatory, Harvard. The largest and brightest, Ceres, was discovered in 1801 by Giuseppe Piazzi, while the next in size—Pallas, Juno, and Vesta— were also found in the first decade of the nineteenth century. After that, Astraea was discovered only thirty eight years later.

In 1891, Max Wolf, mastermind behind the Zeiss planetarium, pioneered the use of astrophotography to detect asteroids. The swiftly moving bodies appeared as streaks on long-exposure photographic plates, and could be picked out far more easily than before. The frequency of detection has increased inexorably since, and more than a hundred thousand are now catalogued. The combined mass of all objects in the main asteroid belt, those between Mars and Jupiter, is still very small—only four per cent of the mass of the Moon. Ceres, at just under a thousand kilometers in diameter, but still just a misty blur when observed even by the Hubble Space Telescope from space, represents one third of the total. The next three most massive asteroids bring the total to just over half, and the number of asteroids then increases rapidly as their individual masses decrease.

Most of the other leftovers from the earliest days of the solar system have been swept up by more massive bodies as time has passed. A spectacular example of this process in action was the impact of Comet Shoemaker–Levy with Jupiter in 1994. Countless small protoplanetary bodies, far out in the cold empty wastes of the embryonic solar system, never collided or never collapsed onto other accreting matter. They have left us with a vast reservoir of small, icy cometary bodies circling far out beyond Pluto in the Oort cloud.

The manner in which the early gaseous disk collapsed means that the planet and asteroid orbits are pretty much confined to a flattened plane, referred to as the ecliptic. Because the Earth's spin axis is itself tilted about this plane, the paths of the planets on the sky also trace out an inclined plane. From our understanding of the laws of gravity we know that the planets circle the Sun with orbital periods which increase with their distance from it: the Earth orbits the Sun in one year; Jupiter, five times farther from the Sun takes nearly twelve years; while Neptune, some thirty times farther from the Sun than the Earth takes more than one hundred and sixty Earth years to complete its distant orbit around the Sun.

OVER THE LAST TWO THOUSAND YEARS, our view of the solar system has been transformed. Plato and his contemporaries upheld a geocentric world view in which a few 'wanderers'—the brightest planets—traversed the night sky in complex and inexplicable epicyclic motions. Four hundred years ago, acceptance of the heliocentric hypothesis, and Newtonian gravity describing planetary motions as simple elliptical orbits with the Sun at one focus, could together explain the motions of the Earth and the other planets. But the more recent discovery of other planets, an abundance of moons, vast numbers of asteroids, and the comets hailing from the vast and distant Oort cloud, has radically altered our detailed view of the solar system.

No longer to be thought of as empty space occupied by a few planets, our current picture is of countless objects encircling our Sun, with some extending far beyond even the distant orbits of Neptune and Pluto. With the origin of these various bodies hidden somehow in the earliest formation of our solar system, our picture is of a few massive and vast numbers of smaller bodies orbiting the Sun in a largely systematic fashion, but with a significant amount of chaotic motion unleashed by gravitational encounters, especially of the smaller bodies, added on top. Our Earth sits within this reasonably benign rotating system, but our immediate vicinity is also peppered from time to time by asteroids or cometary debris crossing our orbit on their own celestial journeys.

Our Sun, in short, is a massive hydrogen furnace surrounded by a captive entourage of debris left over from its formation. Today, our view of other stars in the Galaxy is similarly one of islands of massive sources of nuclear energy each surrounded by their own cohort of planets and other objects, albeit at very different stages in their formation cycle—some only recently born, others long dead.

The Hipparcos star catalogue has set in motion many new investigations into our solar system, and other planetary systems.

THE PROBLEM OF WEIGHING AN ASTEROID, like weighing a star, is a challenge. They vary in density and shape, so that estimates from size alone can be quite unreliable. The mass of any celestial body, including the Sun, planets, and other stars, is only accessible directly through their gravitational effects on other bodies. Asteroids can be weighed by measuring the tiny changes in their orbits when two asteroids pass close to each other. Known orbits are used to predict forthcoming close encounters, and the orbital tracks before and after are measured as carefully as possible.

Forty eight minor planets were bright enough to be contained in the Hipparcos observing programme, and a particularly favourable pair for this particular trick were 20 Massalia and 44 Nysa[48]. A succession of three encounters had been predicted just before or during the Hipparcos observations, with favourably low encounter speeds of around a kilometer per second. An accurate mass for Massalia was estimated from tiny shifts in its orbit. Its mass, and its diameter of a hundred and fifty kilometers, imply a density roughly that of concrete. An improved mass and density for the largest asteroid, Ceres, was found in a similar way after observing its gravitational pull on four other asteroids.

The mass of one individual asteroid is not, in itself, a revolution. But accurate asteroid masses are diabolically difficult to measure. Taken together with their sizes which are more easily measured, we find estimates of their density. It is the density which gives important hints about their composition, structure, and formation. The rotation speed of several asteroids have also been measured from the changes in sunlight reflected off their irregular surfaces. Improved positions of Uranus and Neptune, of Jupiter's satellites Europa, Ganymede and Callisto, and of Saturn's satellites Titan and Iapetus, have all been used in a better description of their motions, and in developing careful models of how bodies move through the solar system.

Occultations of stars, by asteroids or planets passing in front of them, gives knowledge of the shapes and sizes of many solar system objects, and even details of their atmospheres. Improved star positions from Hipparcos greatly improved the ability to predict an occultation, and in turn better planning of observing campaigns to exploit them. Around thirty occultation events are now observed each year, and from them have come the first observations of lunar meteor impacts in November 1999, the occultations of binary asteroids, and occultations of the outer Kuiper belt objects, all giving new information on their shapes and sizes.

The interpretation of some of these events can be surprisingly detailed. Unvisited by spacecraft, Pluto's tenuous nitrogen atmosphere was first detected by occultation in 1985, and studied more extensively by a second event in 1988. These led to the conclusions that Pluto's atmosphere is composed primarily of nitrogen gas in equilibrium with surface nitrogen frost. Observations in July and August 2002, predicted from the Hipparcos star positions, showed a doubling of Pluto's atmospheric pressure over the intervening fourteen years. Even though Pluto was moving further away from the Sun during this part of its elliptical orbit, the evaporation of the south polar nitrogen frost cap was nevertheless increasing as it emerged into sunlight following the winter equinox there in 1987. We were observing the climate changing on Pluto, six billion kilometers away.

WITH CLIMATE CHANGE on Earth high on the current social and political agenda it may seem curious that measuring the positions of other stars can have anything to say about our own weather patterns. But astronomy offers a unique view on certain issues of long-term climate change, and on habitability and how life might be sustained over long periods of time in planetary systems beyond our own.

The Intergovernmental Panel on Climate Change has compiled compelling evidence for significant effects attributable to anthropogenic causes, notably the build up of carbon dioxide in the atmosphere due to the burning of fossil fuels. These I for one do not contest. But effects related to the motions of bodies in the solar system also have an important bearing on long-term climate change over thousands and tens of thousands of years.

Let us start with the Earth itself. Shifting land masses over geological time due to plate tectonics contribute to climatic variations in a variety of ways: in moving the land masses closer to or further from the poles or equator, shifting the global ocean current circulation according to their locations, and changing the mean height of the land through major mountain-building orogenies. Large variations in the chemical composition of the Earth's atmosphere over geological times have occurred, and this and many other phenomena such as volcanic activity all play their role. During the depths of the last ice age 18 000 years ago, when vast glacial ice sheets were stacked high on the land, the sea level was a hundred and twenty meters lower than today, while during the previous interglacial some 120 000 years ago, the sea level was for a short time six meters higher. And compared to the present rise of almost two millimeters a year in the past century mainly as a result of human-induced effects, poorly-understood episodes of meltwater pulses appear in Earth's geological records. One such event around 14 000 years ago led to a global sea-level rise of about twenty five meters over just five hundred years, twenty five times the present annual rate. The current changes are, let us make no mistake, alarming. But so too is the realisation that natural changes of a far greater magnitude have occurred in relatively recent history for reasons that are not at all well understood.

AMONGST THIS BEWILDERING COMPLEXITY, I will focus on some areas where Hipparcos has contributed to a deciphering of Earth's climate as influenced by certain astronomical phenomena.

The simple fact that the Earth's spin axis is tilted with respect to the ecliptic plane, and that the Earth's orbit around the Sun is elliptical, together lead to the seasonal changes: the northern hemisphere is tilted towards the Sun in the northern hemisphere summer, and vice versa. The Tropic of Cancer[49] is the northernmost latitude at which the Sun can appear directly overhead at noon. This excursion peaks at the June solstice, when the northern hemisphere is most tilted towards the Sun. Similar conditions define the Tropic of Capricon in the south. These imaginary 'lines' currently lie a little more than twenty three degrees north and south of the equator.

The detailed path of the Earth's motion actually changes rather significantly over time. The most dominant effects, sometimes referred to as the Milankovitch cycles, include the changing ellipticity of the Earth's orbit, and the changing tilt of its spin axis with respect to the ecliptic plane in which the planets lie. Astronomers split the changing tilt into three parts—a long-term motion called precession, a medium-term motion called nutation, and a shorter-term less-predictable wobble called polar motion. Most of the first two effects are due to the fact that the Earth is somewhat flattened due its rotation. This flattened sphere responds in a more complex way to the gravitational tugs of the Sun, the

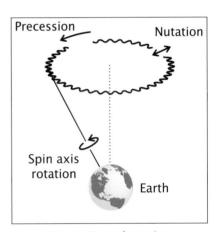

Precession and nutation

Moon, and the other planets in our solar system. The last of these effects, polar motion, is influenced more by the changing conditions within the Earth itself—subtle effects such as circulating ocean currents, wind systems, and ghostly liquid motions deep in the Earth's core.

Combined, these motions lead to variations of the Sun's energy reaching the Earth's surface, over months, over years, and over hundreds and thousands of years. Although the effects are probably insufficient to start or end an ice age, they can explain, in large part, the episodic nature of the Earth's glacial and interglacial periods within a given ice age, including that over the last two million years. For example, the eccentricity of the Earth's orbit, the amount it deviates from strictly circular motion, fluctuates on a cycle of about a hundred thousand years, between zero and five per cent, a variation which is of central importance to the glacial cycles. Currently at about three per cent, the present value leads to a six per cent increase in the solar energy falling on the Earth in January compared to July—hard though this might be for inhabitants of northern climes to accept.

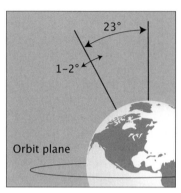

Long term changes in the inclination of Earth's spin axis

The Earth's axis tilt also rocks gently back and forth by more than a degree with a period of just over forty thousand years. Currently at about twenty three degrees, this tilt largely accounts for the yearly cycle of the seasons. Smaller tilts, over thousands of years, give less seasonal contrast in solar illumination, but a greater contrast between the equatorial and polar regions, possibly promoting the growth of ice sheets. Precessional wobbling of the spin axis has a period of slightly less than twenty six thousand years. Climatic variations are then larger when the summer and winter solstices coincide with the times when the Earth, in its elliptical orbit, is at the extremities of its orbital path around the Sun.

Careful and repeated measurements of star positions reveal how the Earth's axis is pointing in space at a given time, and they have provided all of the above information over the last few decades or more. The discovery of precession is generally attributed to the ancient Greek astronomer Hipparchus (circa 150 BCE). Only much later was it explained by Newtonian physics as a consequence of gravitational forces acting on the Earth's equatorial bulge. The largest component of the Earth's nutation has a period of just over eighteen years; it was discovered by James Bradley who announced it in 1748, almost two decades after his discovery of the aberration of starlight. Detailed theories of precession and nutation have been improved successively over the last century, and the Hipparcos observations enter the very latest descriptions adopted by astronomers' governing body, the International Astronomical Union, in 2000.

THE MORE RAPID and less-regular wobbling of the Earth spin axis is called polar motion, predicted by the formidable Swiss mathematician Leonhard Euler in 1755. Based on the known flattening of the Earth, he estimated that it would have a wobbling period of 355 days. Several astronomers searched for evidence of this motion, to no avail. American astronomer Seth Chandler extended the search, looking over a wider range of periods, and observed it finally in 1891. It had a period a little longer than Euler had calculated, at about 433 days. The wobble is truly small: just seven tenths of a second of arc, corresponding to a mere fifteen meters on the Earth's surface.

Once detected, the difference between the observed period and that predicted by Euler was explained by Canadian–American astronomer Simon
Newcomb (1835–1909), director of the Nautical Almanac Office in the USA. Newcomb inferred that it was
caused by the non-rigidity of the inner Earth. This so-
called Chandler wobble combines with another wobble
with a period of one year, to give a total combination
which varies with a period of about seven years.

Simon Newcomb

In this context, a particularly surprising application has been made possible by the Hipparcos positions, allowing us to track the detailed Earth wobble
over the past century. Given that the satellite made its
observations for just three years around 1990, this might sound a little surprising, and this is how it was done.

The discovery of the Chandler wobble more than a century ago precipitated an impressive and coordinated international response to advance its
understanding. From 1900, the International Latitude Service[50], established
by the International Association of Geodesy, set up a network of measurement stations around the world. The idea was that by measuring stars as
they passed overhead, the differences between expected and observed positions gave direct evidence for the wobbling Earth on which the telescopes
themselves were fixed.

Observations to determine the wobble started
at Carloforte, Cincinnati, Gaithersburg, Mizusawa,
Tschardjui and Ukiah, California (where 28-year
old Frank Schlesinger started his astronomy career as the lone astronomer in charge), with more
at Greenwich, Pulkovo and Washington soon to
follow; new participants from the 1930s included
Belgrade, Kitab and Poltava, with Beijing, Grasse,
Moscow, Paris, San Fernando, San Juan, Shanghai,
Tokyo, Yunnan, and others joining the measurements from the 1950s.

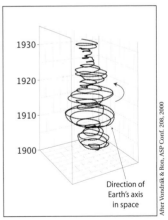

Earth axis wobble, 1900–1930

From 1955, observations of the Earth's wobbling motion were made by between sixty and
ninety observing stations, with all the measurements coordinated by the Bureau International de
l'Heure in Paris. At the peak of the programme in the mid-1970s, two or
three thousand observations a year were being made. These historic measurements from the wobbling Earth provide unique information on the ir-

regularities of the its orientation over the entire twentieth century, although satellite observations and lunar laser ranging have taken over in the last two decades. But the observations themselves had never been fully exploited, because the star positions used as reference points were themselves of restricted accuracy.

Hipparcos stepped in by providing star positions and motions of unprecedented quality. So much so, that we can propagate the star positions observed by the satellite backwards in time, to give a reference catalogue highly accurate throughout the last hundred years. Even for observations made in 1900 for example, Hipparcos gives positions far better than any which existed at that time.

Jan Vondrák (2007)

The mountain of old observations were collected, and re-interpreted according to the star positions which ranged around the heavens on the relevant dates. The piecing together of this work, by Jan Vondrák and colleagues from the Czech Academy of Sciences in Prague, has given us the best picture to date of the Earth's spin axis motion over the last hundred years. With greater clarity than ever before we can see that the wobbling amplitude has varied continuously since its discovery, reaching its largest size in 1910 and again in 1950, and fluctuating noticeably over the decades. The wobble seems to be maintained by changes in the mass distribution of the Earth's outer core, its atmosphere, oceans, and crust. The most recent idea—perhaps the solution to the Chandler Wobble mystery—is that the main driving cause is changing pressure at the ocean bottoms, caused by a combination of temperature and salinity changes, and wind-driven movements in the circulation of the oceans.

How remarkable it is that the miniscule wobbling of the Earth's spin axis from the early 1900s onwards has been characterised by accurate star positions measured almost a century later.

THE VARIATION IN SOLAR ENERGY OUTPUT, and its possible influence on the Earth's climate, has been suspected for more than two hundred years. In 1801, in the Philosophical Transactions of the Royal Society, William Herschel (1738–1822) published his findings of an anti-correlation between the price of wheat and the number of sunspots visible on the Sun: the price went up as the number of sunspots diminished. His work effectively began the study of solar influences on the Earth's weather. While recent man-made contributions to climate change are now largely uncontested, other processes related to solar activity also appear to contribute.

Sunspots are regions of intense magnetic activity, which slow the movement of heat outward from the Sun's interior, and lead to localised areas of reduced surface temperature. They are frequently visible if the Sun's disc is projected onto a plain surface, even without the aid of a telescope. Sunspots appear dark, even though the temperature of these regions is still around 4000 °C. Herschel believed that sunspots were openings in the Sun's luminous atmosphere, an opinion that was to endure well into the nineteenth

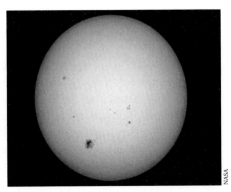

Sun showing sunspots

century. There was even speculation that they hinted at evidence that the Sun was inhabited.

Telescope observations of sunspot numbers, dating back to the time of Galileo, show marked increases and decreases roughly every eleven years, corresponding to a variation in the Sun's surface magnetism. But records also show that in addition to the eleven year cycle, the solar magnetism drops to very low levels for extended periods every few centuries.

Chronicled evidence for significant climate changes on Earth extends back over the last millennium. A particularly cold period which caused considerable hardship was the 'Little Ice Age', from the sixteenth to the mid-nineteenth centuries[51]. Europe, North America, and perhaps much of the rest of the world, were subjected to bitterly cold winters. In the winter of 1709, for example, temperatures in Europe sank below −20°C, and walnut trees froze to death. The coldest excursions apparently coincided with the virtual disappearance of sunspots, the so-called 'Maunder Minimum' of 1645–1715, named after the solar astronomer Edward Maunder (1851–1928). Maunder called attention to this dearth of sunspots after studying historical records, inspired by similar work by Gustav Spörer (1822–1895).

Other cold episodes now recognised include the Spörer Minimum (1460–1550) and, less markedly, the Dalton Minimum (1790–1820). Geological records based on the carbon 14 isotope have been used to extend information on the Sun's magnetic variations back over thousands of years, and yet more ancient minima are now recognised. Contrasting warmer periods have also existed. Most notable amongst these were the 'medieval maximum' around 1000 CE, a particularly clement period accompanied by the migration of the Vikings.

The argument connecting our climate with sunspots runs as follows. We know that the Earth is bathed in a continuous stream of high energy 'cosmic rays' produced elsewhere in our Galaxy, partly responsible for the production of carbon 14. An increased solar wind when the Sun is active reduces the Galactic cosmic ray flux normally reaching Earth. When the Sun is relatively inactive, more carbon 14 is therefore expected. And what is found is that the extended cold periods do coincide with peaks in carbon 14 seen in tree rings. The suggested link is that increased cosmic ray intensity leads to ionisation of tropospheric aerosols, itself leading to the condensation of cloud droplets, and hence to a global lowering of temperature.

If there is a connection between solar activity and climate, and a pronounced drop in the Sun's activity occurs again, we could be plunged into another long and extended cold spell, global warming or not. The uncomfortable reality is that we have no real idea why the Sun shed its sunspots during the Maunder Minimum, and no way of predicting whether or when such a change might occur again.

Jason Wright (2009)

With such a precipitous gap in our understanding, knowing whether this is a common or rare occurrence would be a start. Jason Wright, at Berkeley, assembled a sample of Hipparcos stars out to a distance of two hundred light-years, chosen to be as much like the Sun as possible. He argued that if, say, five per cent were in a very low activity state, then the low state probably also exists in the Sun for around five per cent of the time. The more stars in a low state, the more common the state, and the higher the chances of a new Maunder-type minimum occurring in the future. Monitoring the activity of such stars from the Mount Wilson observatory since 1966, has found some in a high state, others in a low. If a switch between the two is detected in any of these stars in the near future, new light might be thrown on the Maunder mystery. Meanwhile, the jury is still out as to how soon a new cold period might be upon us.

OUR SUN'S TRAVELS AROUND THE GALAXY might also be related to very long-term climate variations. Since its formation about 4.5 billion years ago, the Sun has made about twenty orbits around the Galaxy, once every two hundred and fifty million years. The spiral arms also rotate around the Galaxy, but with a different rotation speed which is not so easy to discern. Depending on their relative rotation speeds, the Sun may have made several passages through the spiral arms in its far distance past. As it travels through a spiral arm, it would be subject to higher rates of cosmic rays due to the bursts of star formation, and their accompanying death throes, going on around it. It seems plausible, therefore, that the Sun's passage through the spiral arms should be accompanied by very extended cold periods. What we see in geological records is evidence for great ice ages around 20, 150, 300, and 450 million years ago. Direct evidence for these great glaciations on Earth are, incidentally, confirmed by the ancient tropical sea temperatures measured through oxygen isotope levels in biochemical sediments.

The Hipparcos observations have now given us a better estimate of the speed with which our Sun moves around the Galaxy. Other Hipparcos studies are providing better estimates of the spiral arm rotation speed. Models now suggest that the great ice ages of many millions of years ago could coincide with our Sun, and the solar system with it, passing through the spiral arms. According to the best estimates today, the Sun passed through the so-called Sagittarius–Carina spiral arm during the Miocene epoch some 20 million years ago, through the Scutum–Crux arm during the Jurassic to early Cretacious periods some 150 million years ago, through the Norma–Cygnus arm during the Carboniferous period some 300 million years ago, and through the Perseus arm during the Ordovician to Silurian periods some 450 million years ago.

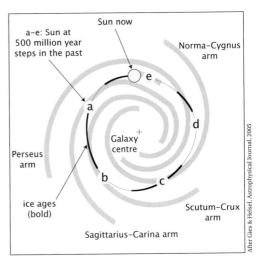

Sun's path through the Galaxy's spiral arms

But here we have some good news too: the next very extended, very cold period on Earth, due to this effect at least, still lies tens of millions of years in the future.

AS STARS MOVE AROUND THE GALAXY, they do so in very different orbits. Some are pretty much circular, while others are more elliptical, rather like those of the planets orbiting our Sun. We can tell from the Hipparcos measures that the orbit of our Sun, and therefore of our solar system, is rather circular—it neither plunges much closer to the centre of our Galaxy than it is now, nor does it travel much further out. It follows a rather privileged path over millions of years, not changing much in its Galactic environment, stable enough perhaps for life to have stood a better chance of gaining a foothold.

But few stars orbit precisely in the mid-plane of the Galaxy disk. If they are born with some up and down motion, or are somehow imparted with some movement a little out of the plane, they will continue see-sawing around, up and down, as they travel around the Galaxy. A little upward motion and they will rise a little, perhaps a parsec or two, before being dragged down again by the combined gravity of the disk, passing through it, and continuing to bob up and down in this way for aeons. A little more upward motion imparted, and they might rise ten or twenty parsecs or more before being tugged back down again, out through the other side of the disk, the larger oscillation continuing again for hundreds of millions of years.

As we have seen, our space measurements tell us that the Sun currently sits at about twenty parsecs above the mid-plane, and that it moves up and down with an oscillation period of about eighty million years—it bobs up and down just three times in the two hundred and fifty million years that it takes to make a full revolution of the Galaxy. Neither does it rise far above or below the plane, a fortuitous circumstance which may have protected life as it evolved over the hundreds of millions of years past.

We have also seen that the prospects of a direct hit with another star are extremely remote; we would need to travel for ten thousand billion years, a thousand times more than the present age of the Universe, and more than a thousand times longer than the Sun itself will live, for this to have reasonably chance of occurring. But this is not quite the full story: head on collisions would be the end, but near misses, and perhaps not so near misses could still be pretty catastrophic. According to these various estimates, the Sun last hurtled through the Galactic mid-plane two or three million years ago and, before that, at regular intervals of about thirty to forty million years, half of its vertical oscillation period. These spaced out traverses through the more crowded star fields of the plane would have been marked by an increased number of nearby star passages.

There is a presumed relation between near-miss encounters and increased meteorite impact events on Earth which runs as follows. The solar system is believed to be surrounded by the Oort cloud, a vast reservoir of billions of comets with orbits extending from thousands of times the Earth's distance from the Sun out to true interstellar distances, and with a total mass of just a few times that of the Earth. The outer regions of the Oort cloud are only loosely bound to the solar system, and easily affected by the sporadic gravitational pull of passing stars. These forces may occasionally dislodge comets, and send them towards the inner solar system, with possible consequences for an increased impact hazard on Earth. These may be the cause of bursts of impact cratering and perhaps mass extinctions that seem to be evident in paleontological records on Earth. We know that these bombardments took place, and there are some subtle hints that the attacks may have been periodic.

The largest-known, the 250 kilometer Vredefort crater in South Africa, was the site of a ten kilometer monster fireball that hit Earth around two million years ago. The slightly smaller Sudbury basin in Ontario was similarly caused not much later in time. Both are, as a result of induced volcanism, sites of concentrated reserves of platinum group metals, including nickel, copper, and gold. Impacts have continued since then, wreaking massive impact craters, including the Morokweng beneath the Kalahari Desert at

Rim of Barringer Crater, Arizona

one hundred and forty five million years before present; the Chicxulub in the Yucatán at sixty five million; the Popigai (Siberia) and Chesapeake Bay events at thirty five million, and the Kara–Kul in the Pamir Mountains of Tajikistan, at a mere five million years ago. All were cataclysmic. The Barringer Crater, Arizona, better preserved and awesome enough at just 1200 meters diameter, resulted from a mere fifty-meter sized boulder hit some 50 000 years ago.

There is, unfortunately, no lack of their brethren, even now stacked up in some inevitable flight path towards Earth. Others will arrive, for certain, with no benefit aside perhaps from a fall in the price of gold. Scientists now studying the Oort cloud and the Kuiper belt, who train their telescopes on comets and asteroids, are advancing pure knowledge, but they are also doing their bit to figure out what events will influence Earth's future destiny.

FINDING OUT WHICH PARTICULAR stars passed near to the Earth in the geologically-recent past, or will do in the future, is simple in principle.

Armed with the details of the Sun's movement in space, and the distances and complete space motions of nearby stars, we can run the movie backwards or forwards in time and follow their relative tracks through space. Their movements point firmly from whence they came, and where they are heading. Joan García Sanchez, a PhD student at the university of Barcelona, working closely with Bob Preston and Paul Weissman at the Jet Propulsion Laboratory in Pasadena, carried out a particularly careful study. The nearest star to the Sun is currently Proxima Centauri, at just over a parsec, and it will come a little nearer with its closest approach to us in about twenty

Joan García Sanchez (2005)

five thousand years. The fastest in angular motion is Barnard's Star, moving through an angle of ten seconds of arc each year, and it will pass us at a distance of just over one parsec slightly less than ten thousand years hence.

Wilhelm Gliese (1983)

ARI archives

But the star to watch out for is Gliese 710, from Wilhelm Gliese's pioneering catalogue of nearby stars. Although still several parsecs distance, it's coming directly at us, and will pass at less than half a parsec in a million years from now. Even so, it will still have a pretty negligible effect on the number of new comets that are thrown into the inner solar system. But these studies remain very incomplete—we'd like to test out the effects of more distant and fainter stars, and trace possible encounters much further back in time.

A slightly better view will also await astronomers studying planets around other stars in just over a million years from now. One of the known exo-planetary systems HD 217107, presently some way from us at nearly twenty parsec, will sail past at just over two parsecs distance.

THE DISCOVERY OF THE FIRST PLANETS beyond our solar system was reported in 1995, and took the astronomical world by storm. Even a year or two before their discovery, discussions of other worlds was almost taboo, the preserve of science fiction writers, not serious scientists. How perceptions have changed in such a short time. Many astronomical institutes now have scientists working in the field, new ways of detecting and characterising them are being developed, space missions devoted to them are being launched, modellers are busily calculating properties of the atmospheres of known systems, and techniques are even now being developed to study their moons.

Most exo-planets, planets around other stars, have been discovered so far using very careful measurements of the host star's radial velocity, the velocity of the star towards or away from us along the line-of-sight, which can be measured using the Doppler technique. A planet going around the star tugs the star back and forth as it moves around its orbit. Given accurate measurements, and enough of them, we can detect the presence of a planet, determine its period, and estimate its mass. Even though the first such planet beyond our solar system was discovered only in 1995, by late 2009 more than four hundred have now been found this way. A few stars even possess multiple planets, in which the Doppler motion of the star is modified by two or planets encircling it. Most planets so far detected are quite big and heavy, much more like our own gas giant Jupiter—itself about three hundred time the mass of the Earth—but some are much lighter.

As the accuracy of the planet-hunting telescopes improves, planets more like the Earth are being found. Some surround the nearby 'Gliese' stars, first catalogued by Wilhelm Gliese, and measured so carefully by Hipparcos. The variety of systems so far detected is bewildering: there are some very massive planets which orbit their star very much closer in than in our own solar system, with objects the size of Jupiter moving within the orbit of Mercury. We still have little idea how these systems originated, or how common systems resembling our own solar system might be. It does look as if at least one in ten stars in our Galaxy may have planets in orbit around them. This opens up an enormous panorama for future studies, with billions of stars in our Galaxy perhaps hosting planets of some form. Given the way we think our solar system came into being, as a collapsing rotating gas cloud, perhaps it is not surprising that planets will be common throughout the Universe.

ANOTHER WAY OF DETECTING PLANETS is to monitor the light of the star as continuously and as carefully as possible, and watch for any dips in the light from the star as the planet passes in front. Anyone who saw the most recent transit of Venus across the face of our Sun in June 2004 will have a good feel for the problems faced. As Venus passed directly between the Sun and Earth, it obscured a small portion of the Sun's disk, appearing from Earth as a small black spot moving across the face of the Sun, and taking just a few hours to pass across it. Prosaic enough when thus described, it was for me a stunning sight, a rare but direct glimpse of an alien world, albeit in silhouette.

This is the type of miniscule and highly improbable conjunction that transit hunters are now tracking down. They've found a couple of dozen examples already, and the numbers are increasing steadily. Most are sentinel planets the size of Jupiter but, quite bizarrely, are often tucked in surprisingly close to their own sun, with periods as short as just a few days. How they got into these close-in orbits currently remains a mystery.

Smaller planets the size of Earth are impossible to see from the Earth's surface, again because of the rippling effects of the atmosphere, but they can be seen from space. A small but pioneering French space telescope, Corot, in orbit since the end of 2006, is now looking for more planet transits from space, while the Hubble Space Telescope has seen a couple of others. Kepler, a much bigger space experiment to extend the census still further was launched by NASA in March 2009. The reason for all the excitement is that by observing the transits at different wavelengths, astronomers can see the presence of specific elements or molecules in the exo-planet's atmosphere. Already, with the acuity of observation from space, this trick has been used to detect hydrogen, oxygen, carbon, and even sodium in the atmosphere of one of the most prominent.

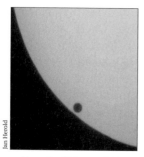

Transit of Venus, 2004

Jan Herold

I stress again that it is not only the positions and motions of stars which fall under the umbrella of astrometry, but also planets and satellites throughout our solar system. Observing their positions, appealing to the theory of gravity, and developing better models to match the continuously improving data, brings us to the point today where our predictions of planetary motions are superbly accurate. From historical records and present-day models, we know that the transits of Venus are among the rarest of predictable astronomical phenomena. They currently occur in a pattern that repeats every 243 years, with pairs of transits eight years apart, separated by desolate gaps of more than a century. The last pair were in December 1874 and December 1882. The transit of 8 June 2004 was the first of the present pair, and its twin will be visible on 6 June 2012. After which, you will have to wait until December 2117 to see this astonishing spectacle of Nature.

Such is the exquisite accuracy in our knowledge of the masses and orbits of solar system bodies—including the Earth—that an even rarer phenomenon can be predicted. A simultaneous occurrence of a transit of Mercury and Venus is possible, but exceedingly rare. According to gurus Jean Meeus and Aldo Vitagliano, such an event last occurred on 22 September 373 173 BCE, and will next occur on 26 July 69 163, and thereafter only in March 224 508, just shy of a quarter of a million years from now. We can be rather confident that nobody has ever seen this absurdly rare double transit, and we may wonder whether anybody ever will. Martians, incidentally, will see a simultaneous transit of Venus and Earth in the year 571 741.

RETURNING TO EXO-PLANETS, astrometric measurements have a particular relevance. We saw that repeated measurements of a star lead to changes in position due to its proper motion and due to its parallax. If the star is a binary, even more complex motions of the centre of gravity of the star light can be measured, giving information on the period and radius of the orbit. If the measurements are accurate enough, the same measurements can be made for exo-planets. In the same way that repeated Doppler velocity measurements reveal the backward and forward motion of the star, and hence the planet in orbit around it, repeated astrometric measurements reveal the side-to-side component of the same orbital motion. The astrometric method has an important advantage: it's possible to measure the planet's mass directly, and not just a lower limit. For multiple planetary systems it's also possible to measure the relative inclination of the planetary orbits—to establish if they all moving in the same plane, as they do in our solar system, or not.

For a number of stars on the observing lists of the ground-based radial velocity planet hunters, the Hipparcos observations gave a useful handle on their true masses. Two of the known transiting planets were found to show up in the intensity data from Hipparcos and, because of the fifteen year time difference between the two types of observation, the orbits could be pinned down with remarkable precision. For the star–planet system HD 189733 the orbital period is just over two days, pinned down with an accuracy of about one second. Studying this star with an array of optical telescopes on ground working together gave an angular diameter of a little more than one third of millisecond of arc. Combined with the parallax distance gave a star diameter of a little less than that of our Sun, and a planet radius just ten per cent more than that of Jupiter.

JUST AS ASTRONOMERS NEARLY two hundred years ago focused their own search for parallax by hunting down proxies for proximity, so astrobiologists and planetary scientists are now urgently trying to narrow down which of the known exo-planets hold the best prospects of harbouring life. Where should we start a search for other habitable planets? Where should we start a search for other life? It is surely a mistake to believe that life elsewhere in the Universe might only take hold under conditions similar to those on Earth, but the Universe is an awfully large place to search blind, and narrowing down the field somehow makes much sense.

In our immediate Galactic precinct, there are around two hundred and fifty stars known out to a distance of about ten parsecs (or thirty light-years), and about two thousand out to twenty five parsecs. We can extrapolate these numbers outwards in space, and infer that there are probably thirty thousand or more stars out to about a hundred parsecs. The stars in such a volume of space are of a staggering variety—of masses and sizes, of temperatures and luminosities, of chemical composition and age, of magnetic activity and rotation speed.

We have seen that our knowledge of the inner structure and of the nuclear reactions that go on within the stars is now in rather good shape. Many of the detailed properties of stars can be well modeled by what are now highly elaborate computer simulations. We can use this kind of knowledge to search for other stars like the Sun, stars which might have planetary systems like our own.

We know that the Sun does rather well in its 'Goldilocks' properties—it's not too hot and not too cold, not too young and not too old. Many other conditions seem favourable also, and these kinds of arguments have focused on a search for solar 'twins'. Any stars which are identical to the Sun in all these respects might naturally be considered as being those most likely to possess planetary systems similar to our own. It's not too much of a leap of faith to argue that they would also be well-suited to host life forms based on carbon chemistry and water oceans.

Hipparcos provides enough information to narrow down the search considerably, because we can sift through the hundred thousand stars now catalogued, and come up with a list of their sizes, temperatures, masses and ages, partly relying on our computer models. These studies have picked out the star 18 Scorpii, at about fifty light-years distance, as most like our Sun. A further handful seem to be reasonably similar in many of their properties, including both stars of the nearby binary pair 16 Cygni. If we accept that a precise solar twin is an unnecessarily tight restriction, studies based on distances and orbits around the Galaxy throw up thirty or so stars within ten parsecs which seem like a good place to start a search. Many of these are already on the list of stars with known planets around them. They are being ever more closely followed by the planet-hunting telescopes on Hawaii, at the Lick Observatory in the USA, at the Anglo–Australian telescope in Australia, and from the European Southern Observatory's mountain top observatory of La Silla in Chile.

SOLAR TWINS LOOK just like the Sun does today. We have seen that all stars, including the Sun, evolve with time. The Sun was somewhat different a hundred million years ago, and will be somewhat different again a hundred million years in the future. We can use our improved knowledge of nearby stars, and our prescriptions of how we believe that stars evolve over time, to search for those most likely to have looked, at some time in their recent past, or those that will look, at some time in their fairly near future, very similar to the Sun. Studying these solar 'analogues' therefore effectively provides a glimpse of our Sun at other points along its evolutionary history.

Klaus Fuhrmann (2008)

HD 63433 is a star originally listed in the substantial Henry Draper catalogue of around 1920 which has a mass, temperature and chemistry very close to that of the Sun. Klaus Fuhrmann, at the Max Planck Institute in Munich, established its age as only two hundred million years, and it provides us with the best example of what the Sun might have looked like soon after its birth, around 4.5 billion years ago. Perhaps it is a star to study in detail to discern conditions on Earth all that time ago. At just over sixty light-years distance, it's on a celestial trajectory coming more-or-less directly at us, expected to sail past our solar system, just six light-years distant, in just over a million years hence. It will provide our descendants with an excellent prospect of studying the young Sun in detail. Other researchers have found Beta Hydri and HIP 71813 to be two other nearby stars which closely resemble the way our Sun might appear two to three billion years in the future, half way in time from now to its ultimate and inevitable demise.

In studies of the existence of life elsewhere in the Universe, there is a general consensus as to where best to start looking: the host star should not be too massive or it will evolve and die too rapidly for life to develop. Not massive enough and it will have too much X-ray activity. Any life-bearing planet around it cannot be too massive either or it will accrete hydrogen gas just like Jupiter. Perhaps some life form exists deep within these hydrogen seas, but it wouldn't be the easiest place to start a search. If the planet is not massive enough, plate tectonic motions will not be sustained, and carbon sinks will not exist, which may or may not be a pre-requisite. The planet must not be too close to its host star or it will be too hot for liquid water to exist, while too far away it will be a perpetually glacial world.

Although liquid water may not be a mandatory condition for life to develop, searches are being focused on the habitable zone in which water exists in liquid form. In our solar system, this habitable zone is not very wide, perhaps stretching from a few per cent closer to the Sun than the Earth, and a few per cent beyond its present orbit. The Earth, in fact, sits close to its own optimum 'Goldilocks' position, not too hot and not too cold. But an even more stringent condition seems necessary: life took a billion years or more to develop on Earth, and throughout that time its orbit sat within a still narrower range of distance over which liquid water existed for a billion years or more, even as the Sun slowly changed in brightness as its inner fuel was consumed. If we accept that life elsewhere needs liquid water, and if it took as long to develop elsewhere as it did on Earth, then we need to find a star with a planet which has sat within its own continuously habitable zone over a billion years or more. Our knowledge of a given star's luminosity based on its distance, and our detailed knowledge of how the star has evolved in the past based on our understanding of stellar evolution, together allow us to pinpoint which stars are likely to host habitable planets, and which planets are likely to have been endowed with liquid water for the billion years or so necessary for life to have developed.

Planet hunters now keep the Hipparcos catalogue close to hand. It is their *vade mecum*, allowing them to figure out, simply and quickly, just how closely each new exo-planet host resembles the Sun, and just where the planet stands, according to present knowledge, on the list of those most likely to harbour life. The star distances lead to a knowledge of the properties of the host star, its luminosity, and the age of the star–planet system. The planet's orbit can be plotted, and this determines whether it is moving within the habitable zone in which liquid water might be present. And from the proper motion follows the nature of *their* solar system's orbit around the Galaxy, a further hint as to its suitability for life.

I WILL SKIP OVER MANY OTHER AMBITIOUS PLANS that are now underway to zero in on planets most likely to be favourable for astrobiological investigations, and the many avenues being explored to understand what combination of atmospheric molecules are likely to indicate the presence of life. In practice, not too much in the way of tangible results are yet in. But SETI, the Search for Extra-Terrestrial Intelligence, is another area that is also making strides as a direct result of the Hipparcos catalogue.

SETI is a collective name for a number of activities aiming to detect intelligent extraterrestrial life. There are great challenges in searching the sky for a transmission that could be characterised as intelligent, since its direction, spectrum and method of communication are all unknown. One of the most technically advanced searches now ongoing is being led by the SETI Institute of Mountain View, California. This grew out of a NASA funded programme in 1992, using radio dishes and targeting eight hundred specific nearby stars. The programme was[52] "strongly ridiculed by US Congress and cancelled a year after its start." But in 1995 the SETI Institute resurrected the work under the name of Project Phoenix, backed by private funding, and aiming to study roughly a thousand nearby Sun-like stars.

Seth Shostak, SETI Institute

SETI's Allen Telescope Array

The institute is now developing a remarkable, specialised radio telescope array in northern California, named the Allen Telescope Array after the project's benefactor Paul Allen, the entrepreneur who co-founded Microsoft with Bill Gates. The first part of the array started listening in for whispers from deep space in October 2007 with 42 six-meter antennae. Completion of the full 350 dishes will depend on funding, and technical results from this pioneering prototype.

For their new search programme, the institute created their list of nearby habitable systems directly from the Hipparcos catalogue using the information on distances for signal propagation issues, variability for climate stability, multiplicity for orbit stability, space velocity as an indicator of the probably chemistry, and spectral classification as a criterion for various habitability conditions. Combined with theoretical studies on habitable zones, stellar evolution, and three-body orbit stability, these data were used to remove unsuitable stars from the list of plausible targets, leaving a residue of stars that, according to present knowledge, are potentially hosts for complex life forms. The analysis resulted in more than seventeen thousand possibly habitable stars near the Sun, of which three quarters lie within a hundred and forty parsecs. Searches for extraterrestrial intelligence are at this moment ongoing in earnest, based directly on the Hipparcos results.

A surprising application, published in mainstream astronomical journals, is the use of the accurate Hipparcos distances for a specific type of SETI search, motivated by the following problem. Communication between civilisations over large interstellar distances is clearly most efficient when transmitting and receiving a highly beamed signal sent out as a high-power burst, rather than as a continuously propagated transmission spread over the whole sky. For such a scheme to be feasible, the transmitter and the recipient, both of which are unknown to the other, must find a strategy that will enable the transmitter to transmit, and the receiver to receive, at an appropriate time and in an appropriate direction. It's like one of those logic problems that, at first glance, looks unlikely to have a solution.

But an elegant answer does indeed exist, set out by Robin Corbet of NASA's Goddard Space Flight Center following his study of the Hipparcos catalogue in 1999. A very precise synchronisation can be achieved by making use of a class of bizarre natural astronomical event—the extremely powerful but very short-duration gamma-ray bursts. These bursts are short sharp flashes of gamma rays, lasting for just a second or two, and appearing from unpredictable places in deep space at random times. First detected in the late 1960s by satellites built to detect possible gamma-radiation pulses emitted by nuclear weapon tests in space, they were kept under close wraps for years while their nature was debated; the first scientific analysis was published only in 1973.

After decades of study, and a series of special space telescopes designed to probe their nature, they are now believed to signal the collapse of the core of a rapidly rotating, high-mass star into a black hole, or the final collision of neutron stars orbiting in a binary system. Pouring out prolific bursts of energy in a cataclysmic death throe, these rare cosmic firework displays take place billions of light years beyond our own Galaxy. Each day, one or two are detectable with specialised gamma-ray detectors in space. Because they disappear within seconds, an international rapid response system is now in place to register the burst, and to transmit details to optical telescopes around the world for immediate follow-up observations, all helping to understand their detailed nature.

Using a particular gamma-ray burst as the timing and direction marker as a form of a celestial lighthouse, the search for intelligent signals carried out on Earth is concentrated in the direction of a candidate habitable star which has been selected to lie closely aligned, to within a certain small angle, with a specific gamma-ray burst event. The signaling event despatched by an alien civilisation has meanwhile been transmitted in a direction roughly anti-aligned with the same celestial event, such that the gamma-ray burst, the alien civilisation, and the Earth, are roughly co-aligned.

Knowing the distance to the target planet-hosting star, which we do from Hipparcos, allows the time lapse between the gamma-ray burst event and the signal event from our extraterrestrial civilisation to be estimated from simple trigonometry. A planet-hosting star at twenty parsecs from Earth, and lying one degree from the gamma-ray burst event seen from Earth, will transmit an event which will be detected on Earth more than three days later, but one whose arrival time on Earth can be predicted with a precision of better than two hours, greatly simplifying the detection problem.

PURSUED BY SOME with passionate conviction, a research subject viewed by others with disdain, SETI is perhaps a topic roughly where planet detection was in the early 1990s, but with the odds stacked far higher against it. The more so since I happen to believe that, as intelligent life goes, we are quite alone in the unsettling vastness of the Universe. But if a definitive extraterrestrial signal were to be detected tomorrow—admittedly unlikely but which we most certainly cannot exclude—the stampede to join the race can barely be imagined.

At a round-table discussion on the subject of habitable planets with the then UK Minister of State for Science and Innovation, Malcolm Wicks, deep in the bowels of the Department of Trade and Industry in London in June 2007, I made this point. Journalists present picked it up, and Roger Highfield, then with the *The Daily Telegraph*, promptly challenged the other scientific authorities there assembled to corroborate my position, which they did. Thus armed, he turned the considerable authority vested in him to face the establishment: "For heavens sake, Minister, get off your backside and fund this." It was bowled full toss with a touch of humour, and batted away accordingly, but food for thought as we departed our separate ways.

AND IF ANOTHER civilisation is detected, I wondered, who would speak for planet Earth?

SO WE REACH THE END of our tour of history's greatest star map, and but a sprinkling of the state of knowledge of our Universe derived from it at the start of the third millennium.

Building on improvements in measuring star positions over the past two thousand years, we live at a time when the possibilities of space exploration came to the rescue of progress which had largely stagnated under the flickering shroud of our Earth's atmosphere. The satellite was a carefully designed package of high-technology components put together by some of Europe's most advanced technology companies. The catalogue was pieced together on the ground from a vast jigsaw of precision star measurements made from the dark and isolated vacuum of space. The entire venture took thirty years, from the first concept in 1967 to completion in 1997. New scientific insights have followed, pried from the distribution and motion of the stars by curious minds. It is not difficult to imagine that vastly more profound insights into our Universe will follow as measurement accuracies are pushed still further in years to come.

IF THERE IS ONE PARTICULAR mystery that the study of the Universe brings home to me it is the one that sits midway between its origin, and the emergence of life. For the former, we seem forced to conclude that matter magically sprang into existence from nothing at all. For the latter, life somehow emerged triumphantly perhaps from a chance arrangement of molecules from which evolution was thereafter set in motion. Preposterous though both of these pivotal events were, even if we invoke divine creation, each may have demanded just one masterstroke of cosmic connivance.

Astronomy confronts the remarkable spectacle of what happened in between. And it inspires us to ask: what allowed the fundamental atomic particles to have such characteristics that they would duly cast themselves into the hundred or so distinct elements, whose properties are so varied that they would support so many different physical and chemical and biological processes, and in turn be assembled into so many startlingly different molecules, resulting in the rich diversity of the world around us and its plethora of intricate interdependencies? Why should it be that stars could form and burn for aeons, assembling themselves into the vast range of structures which propagate their own existence, churning out the heavier elements, and creating havens of planets of such dazzling variety and inconceivable complexity that the stunning beauty of our living world has emerged and thrived?

I can offer no convincing answers, but my professional life as an astronomer has brought me face to face with some of these questions.

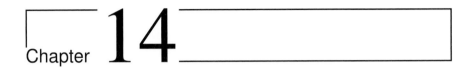

Chapter 14

The Future

If you remained a pupil, you served your teacher badly.

Friedrich Nietzsche (1844–1900)

H ISTORY HAS NOT COME TO AN END. Already by 1997, as the Hipparcos catalogue was being lodged in scientific libraries around the world, astronomers across Europe were pooling ideas for a yet more ambitious experiment to map the stars from space.

This next instrumental leap forward, Gaia, will follow similar principles, but with both scientific ambition and the experiment itself scaled up to reflect twenty years of progress in astronomy and technology, leapfrogging the Hipparcos accuracy by a factor one hundred. It will feature a far bigger lightweight telescope, built from the highly stable ceramic silicon carbide. Like a massive digital video camera, a carpet of CCD silicon sensors almost a square meter in area will record the millions of star images that pass across it as a new orbiting satellite once more scans the heavens.

A powerful on-board processor will handle a vast cascade of image manipulations before the information stream is despatched to Earth. Its data rate will be more than a hundred times that of its predecessor. The satellite's orbit will be rather different—far from Earth, one and half million kilometers away, out along a line from the Sun to the Earth and beyond. Here, at what is termed the Sun–Earth Lagrange point, the combined gravitational and centrifugal forces of these celestial bodies will impose upon the satellite a constant and favourable alignment with respect to the Sun and Moon.

M. Perryman, *The Making of History's Greatest Star Map*, Astronomers' Universe
DOI 10.1007/978-3-642-11602-5_15, © Springer-Verlag Berlin Heidelberg 2010

This next leap in ambition promises a scientific harvest which dwarfs that of Hipparcos. Its colossal survey of more than a thousand million suns will provide a defining census of around one per cent of our Galaxy's entire stellar population, pin-pointing them in space right across its vast expanses. Unimaginable numbers of stellar motions will reveal many more details of the vastly complex motions at play within our Galaxy. It will provide insights ranging from new tests of general relativity to stringent limits on the variation of fundamental physical constants. Planets circling other stars will appear in their thousands from their tiny wobbling motions, identifying candidate systems for the burgeoning discipline of exo-biology.

Rusty Schweickart (2006)

Tens of thousands of asteroids will be measured. Objects which may approach Earth in the coming century will have their trajectories plotted, and if they have our name written on their projected orbit, we will perhaps have time enough to see whether mankind's innate resourcefulness can do something to avert an impending and potentially calamitous collision.

To discuss Gaia's contribution to this emotive challenge, a distinguished visitor to my office in 2002 was the Apollo 9 astronaut Rusty Schweickart. For several years he has chaired the board of the B612 Foundation, championing the development of a spaceflight concept to protect the Earth from future asteroid impacts. As to if, or when, or just how hard we will be hit, his views could not be expressed more clearly: "We're sitting in a shooting gallery, with hundreds of thousands of these things whizzing around in the inner solar system," he has testified to US Congress. "So it's just a matter of time."

Just before that, in February 2001, Lord David Sainsbury, as the UK Minister for Science, reported on the work of a task force comprising Dr Harry Atkinson, Professor David Williams and Sir Crispin Tickell, one-time British representative to the United Nations. These scientific and political heavyweights came up with a package of measures to mitigate the potential threat from asteroids and other near Earth objects in which Gaia would play a prominent role. "The potential threat of asteroids and other near-Earth objects to our planet is an international problem requiring international action", they said.

Far from being a research project of purely abstract relevance, by contributing to an understanding of Earth impact hazards, long-term climate change, and extrasolar planets, this next leap in measuring star positions will see astronomy contributing to an evermore profound awareness of the deep mysteries of Nature.

AFTER FIVE YEARS of studies, and after protracted discussion and intense lobby, the European Space Agency's advisory bodies signed up to Gaia in October 2000, twenty years after a very different body of scientists did the same for Hipparcos in 1980. It targets measurements of ten *millionths* of a second of arc for the brightest stars, a hundred times better than the pioneering results obtained from space by Hipparcos, the width of human hair viewed from a thousand kilometers. It is due for launch from Kourou in 2012, just over two decades following the footsteps of Hipparcos. After a five year programme of scanning the skies at the start of the third millennium, its harvest will be in scientific hands in 2020.

In 2001, Malcolm Longair, then head of Cambridge University's Cavendish Laboratory, made the observation[53]: "It would be rash to guess how astronomy, astrophysics, and cosmology will develop. The key roles of advanced technology and the undertaking of large-scale systematic studies remain paramount. My reason for emphasizing precision space astrometry is that, in some people's eyes, it might not have the 'gee-whiz' attraction of the physics of black holes, of the first microsecond of the universe, or of the discovery of life on other planets. No doubt these are more marketable commodities, and they present wonderful scientific challenges. But, we need to be aware of our scientific roots. We now know how to carry out astrometric programmes from space with one thousand times the precision of Hipparcos, and I firmly believe this programme must be carried out during the twenty first century."

By 2030, it will fall to somebody to try to summarise its achievements after a decade of scientific scrutiny. Perhaps, by that time, some ingenious scientists and space engineers will have figured out how to build a satellite to measure at the levels of a thousand times better than Gaia, at the billionth of a second of arc. At that point, distances out across the vast uncharted expanses of the Universe could be measured directly.

For now, such a possibility remains firmly in the realms of science fiction. Indeed, as undisputed authority Erik Høg has written after his lifelong contributions to the field, and based on his recent studies of its historical and technical development: "The Gaia astrometric survey of a thousand million stars cannot be surpassed in completeness and accuracy within the next forty or fifty years."

History is littered with erroneous predictions, so many self-proclaimed seers consistently failing to anticipate the accelerating pace of change. We will see, in due course, whether this one stands the test of time. It would take a brave person to wager a significant sum either way, but my tendency would be to side with Erik Høg.

I, AND MANY OTHERS, had already worked on the Gaia project for five years by the time of its acceptance in 2000. A colleague in ESA, Giacomo Cavallo, secretary of the Science Programme Committee, an astrophysicist by training but more of an urbane philosopher by persuasion, and one who had done much over the years to shepherd many delicate issues through ESA's political minefield, asked me a few weeks later what I really thought of Gaia. I was a little confused by the thrust of the question, given my deep involvement, and I asked what he meant by it. "Well," he said in thoughtful reflection, "you worked for twenty years on Hipparcos and, far from being some long-lasting contribution to science, its impact will now soon be overtaken by even more remarkable results from Gaia. Isn't that prospect, well, rather devastating?"

I was silent for a moment, because I'd never viewed the progress of science in those terms. We had finished one experiment, and moved swiftly on to the next, for an astonishing opportunity had again fallen into place to do it even better. Some important questions in astronomy had been addressed, but we would be handing down to the next generation the chance to answer so many more. We had made the tools of astrometry available to contemporary astrophysicists ranged across a diversity of research fields.

Most importantly, we had propelled this ancient art into the infinities of space, drawing aside the curtains of Earth's atmosphere, showing how it could be done from far above its surface, and setting out on another long and difficult journey to do it much better. We had teams of motivated, knowledgeable scientists trained up, and a high-technology industry impatient to respond to the challenge. Science, probably, will never come to an end. New questions invite more demanding observations, and new observations inspire more sophisticated theories. Mankind's curiosity has proven to be utterly insatiable. "I feel very excited about it," I said.

BUT AS WE PARTED the words of Charles Wynne twenty years earlier came to my mind, and I thought to myself: "But I do know that milliseconds of arc won't split into microseconds of arc very easily."

Epilogue

THE HIPPARCOS SATELLITE DELIVERED all, and more, that had been asked of it at the time of its acceptance by the international scientific advisory committees of the European Space Agency in 1980. Commissioned to deliver a catalogue of a hundred thousand stars with an accuracy of two thousandths of a second of arc, its final harvest was nearly one hundred and twenty thousand stars a factor of two better, and an additional two and a half million plucked from its star mappers. During the ten years after publication of the catalogues in 1997, several thousand papers based on the results have appeared in the scientific literature. The information contained in the catalogues will continue as a bedrock of astronomy far into the future.

PIERRE LACROUTE, AT THE ORIGINS of a space astrometry mission, died in 1993 aged 86. He had seen the satellite launched and functioning almost to the end of its operational life. Although he did not live to hold the final catalogue, he knew that Hipparcos had been a resounding success.

Project manager Franco Emiliani moved from the project in 1986 to head ESA's Columbus department in Paris. Hamid Hassan, who subsequently led the technical side of the project until launch, went on to manage ESA's Huygens probe of the NASA/ESA Cassini mission to Saturn; he died shortly before Huygens made its spectacular parachute descent to the surface of Titan in January 2005. A reunion of the ESA project team in October 2007 was attended by twenty of the ESA Hipparcos project team members.

Catherine Turon, Jean Kovalevsky, Lennart Lindegren, and Erik Høg, leaders of the scientific teams that worked on the project, were presented with the Director of Science medal at a ceremony of ESA's Science Programme Committee in Bern in May 1999. "As team leaders," enthused Roger–Maurice Bonnet, "our medalists were responsible for the largest computing task in the history of astronomy. ESA says thank you to them and the many other scientists who devoted twenty or thirty years of their working lives to making Hipparcos a success."

M. Perryman, *The Making of History's Greatest Star Map*, Astronomers' Universe
DOI 10.1007/978-3-642-11602-5_16, © Springer-Verlag Berlin Heidelberg 2010

In their own countries, to mark their own pivotal roles in the construction of history's greatest star map, Jean Kovalevsky was elected to the French Académie des Sciences in 1989, and Catherine Turon received the silver medal of the French National Centre for Scientific Research in 1991. Lennart Lindegren received the Letterstedt Prize of the Royal Swedish Academy of Sciences in 1990, and was elected to the Swedish Academy in 2009. The Russian Academy of Sciences awarded Erik Høg its Struve Medal at a ceremony in Saint Petersburg in 2009.

Others have written to me down the years to acknowledge the success of the project after its inauspicious start. Charles Bigot, head of Arianespace at the time of launch, sent his own congratulations for the "spectacular advance" it brought to astronomy, conscious and proud of the key role that his organisation had played in this new understanding of the Universe.

A NUMBER OF SATELLITE COMPONENTS designed and built as prototypes or flight spares remain extant. The full satellite engineering model, resplendant with solar arrays deployed, occupies pride of place in ESA's public space exposition in Noordwijk. Flight spares of the spherical mirror, the tiny 'modulating grid' at the heart of the experiment, and the focal plane, have been consigned into the long-term archives of the Paris Observatory in France, the scene of so many Hipparcos scientific meetings over the years.

The engineering model of the unique 'beam-combining mirror', a mere thirty centimeters in diameter but the most crucial and the most technically challenging piece of the entire instrument, is on long-term loan to the Royal Greenwich Observatory. At the time of writing it is on public display at the National Maritime Museum in London, alongside historical landmarks such as the astronomical instruments of William Herschel and the sea-going clocks of John Harrison, the latest in mankind's long line of ingenious technological attempts to unlock the mysteries of the Universe.

THE HIPPARCOS SATELLITE ITSELF is also on permanent public display. It remains trapped in its sweepingly elliptical geostationary transfer track, hurtling around the Earth in its ten hour orbit between five hundred and thirty six thousand kilometers above its surface. It is visible, if only as a point source of light reflected by the Sun, for those suitably equipped with a small telescope and appropriate celestial coordinates. Depleted of fuel and power, any ability to communicate with Earth was long since lost. No more than a lump of metal, glass and inert electronics, it is tracked—as is all other debris encircling the Earth—by the Inter-Agency Space Debris Committee.

It will continue circling Earth in this orbit, still firmly attached to its unexploded apogee motor, for more than fifty thousand years.

Notes

[1] Before use of the Euro in 1999, ESA used first an Accounting Unit and later the European Currency Unit to simulate the basket of currencies of its member states. To simplify numbers, and ignoring inflation over the intervening period, I label amounts simply in Euros

[2] Accepted practice in astronomy capitalises the word Galaxy when it refers to our own

[3] Billion is used throughout to signify one thousand million, or equivalently 1 000 000 000, or 10^9 in scientific notation

[4] *The Impact of Science on Society* by Bertrand Russell (Routledge, 1985)

[5] A similar analogy was used by Martin Rees in his Oort Lecture at Leiden University in 1999, but it may have been used by others before him

[6] BCE, or Before the Common Era, is synonymous with BC. CE, or Common Era, is synonymous with AD, and is dropped where confusion is unlikely

[7] *The Expanding Universe* by Arthur Eddington (Penguin, 1932)

[8] To commemorate the four hundredth anniversary of Galileo's first recorded observations of the night sky using a telescope, the United Nations scheduled 2009 to be the International Year of Astronomy

[9] The names of the fifty or so brightest stars carry a mix of Greek (e.g. Procyon), Latin (e.g. Polaris), or Arabic (e.g. Deneb) etymology. Extending the naming to more stars, Johan Bayer's 1603 *Uranometria* comprised forty nine constellation maps, and a new ordering of star names in which Greek letters, and thereafter Roman letters, along with the Latin possessive form of the constellation name, were assigned depending on magnitude and location. John Flamsteed's 1729 *Atlas Coelestis* embraced the naming of more stars by assigning Arabic numerals within the constellations from west to east. For fainter stars, a plethora of other names abound, generally their sequence number in one or other observational catalogues, such as the HD (for Henry Draper) or HIP (for Hipparcos) catalogues

[10] The division of the sky into constellations dates back to antiquity. In the western world, the constellations of the northern hemisphere are based on star patterns described by

M. Perryman, *The Making of History's Greatest Star Map*, Astronomers' Universe
DOI 10.1007/978-3-642-11602-5, © Springer-Verlag Berlin Heidelberg 2010

the ancient Greeks. Many for the southern hemisphere were introduced by the French astronomer Nicolas de Lacaille (1713–1762). The constellations are nothing more than widely spaced patterns of bright stars that appear loosely related to each other from the perspective of the Earth, generally with little scientific relevance, and typically bearing little resemblance to the objects they pretend to represent. For centuries the boundaries were arbitrary and somewhat fluid. The 88 constellations now defined by the International Astronomical Union had their boundaries drawn up by Eugène Delporte in 1930

[11] Quoted in *Star-Names and Their Meanings* by R.H. Allen (G.E. Stechert, 1899)

[12] Apparent magnitude is a measure of a star's brightness as seen by an observer on Earth, i.e. as an object appears in the sky taking no account of its distance. It has its origin in the Hellenistic practice (generally believed to have been introduced by Hipparchus and popularised by Ptolemy) of dividing stars visible to the naked eye into six magnitudes: the brightest stars of first magnitude, while the faintest visible to the eye were of sixth magnitude. The system is now more formalised, based on a logarithmic scaling which allows both for fainter and brighter objects. Thus Sirius, the brightest star in the sky, has an apparent magnitude of -1.4, while Venus and Jupiter can reach -4. The full Moon is about magnitude -13 and the Sun is around -26. Once we know an object's distance, we can calculate its 'absolute magnitude', i.e. its luminosity, or energy output

[13] If our home had been Jupiter instead of Earth, parallax angles would be five times larger; but we would have to wait twelve Earth years to trace out a full path around the Sun

[14] Astronomical distance measurements in meters or kilometers, or even millions of kilometers, are unwieldy, although I use them in a few places to make a point. The two most convenient and widely-used units of distance are the light-year and the parsec. The light-year is the distance travelled by light in one year; it's a little less than 10 million million kilometers. The parsec is tied to the geometry of the Earth's motion around the Sun; it's a little more than 30 million million kilometers. I have tended to use parsec when emphasising the measurements, and light-year—which still seems to me a little more poetic—when emphasising their immensity. Multiply distances in parsec by about three, or 3.26 to be more precise, if you prefer distances in light-years

[15] More background can be found in *The Cambridge Illustrated History of Astronomy* by Michael Hoskin (Cambridge University Press, 1997)

[16] *The Rise of Scientific Europe 1500–1800* by David Goodman & Colin Russell (Hodder & Stoughton, 1991)

[17] *Aristarchus of Samos* by Thomas Heath (Oxford University Press, 1913)

[18] The story is told in the popular account *Longitude* by Dava Sobel (Fourth Estate, 1996)

[19] Further details are given in *Dividing the Circle* by Allan Chapman (Ellis Horwood, 1990)

[20] *Road to Riches, or the Wealth of Man* by Peter Jay (Weidenfeld & Nicolson, 2000)

[21] *Star-Names and Their Meanings* by R.H. Allen (G.E. Stechert, 1899)

[22] The publication of 1801 gave only the observations carried out over ten years at the École Militaire. The calculations to produce the catalogue were only made in 1837. It was published in 1847 as 'A Catalogue of those Stars in the Histoire Céleste Française, reduced at

the expense of the British Association for the Advancement of Science, and printed at the expense of Her Majesty's Government'

[23] *Parallax: the Race to Measure the Cosmos* by Alan Hirshfeld (W.H. Freeman, 2001)

[24] More than a century later, Lick observatory astronomer Carl Wirtanen is said to have kept a black widow spider in his office to supply his own instrument needs

[25] Quoted in *The Cambridge Illustrated History of Astronomy* by Michael Hoskin (Cambridge University Press, 1997)

[26] *The Physical Universe* by Frank Shu (University Science Books, California, 1982)

[27] A highly readable account was published by the Savilian Professor of Astronomy at Oxford, H.H. Turner, as *The Great Star Map* (John Murray, 1912). The book is a mine of information, with costs, weight of the plates and catalogue, and many other details

[28] Adriaan Blaauw recalls that Pieter van Rhijn (1886–1960), Kapteyn's successor as director of the Astronomical Institute in Groningen and who Blaauw himself knew well, had told him that Kapteyn had numerical computations of star coordinates carried out by prisoners in Groningen. According to Blaauw: "A number of these tables still exist and are now part of the Kapteyn legacy collection kept in the Groningen University Library where they can be consulted. They are a marvel of neatness and accuracy. The people who made them must have taken great pride in delivering them and one can imagine that it must have given them great satisfaction to contribute in this way to Kapteyn's scientific work." Doubts were raised about the role of prisoners at the Kapteyn Legacy Symposium in 2000, there being no written documentation, but Blaauw vouches for the story's pedigree

[29] *A History of the European Space Agency, Volume 2 (1958–1987)* by J. Krige, A. Russo & L. Sebesta (ESA Publications, 2000)

[30] ESA's Halley comet probe was named after the Italian artist, Giotto di Bondone (1266–1337). Inspired by the reappearance of Comet Halley in 1301, Giotto transformed the Star of Bethlehem into a golden comet in his 1304 fresco *Adoration of the Magi*

[31] 23rd Meeting of ESA's Science Programme Committee, 4–5 March 1980

[32] 24th Meeting of ESA's Science Programme Committee, 8–9 July 1980

[33] Contributions to the History of Astrometry Number 6: *Miraculous Approval of Hipparcos in 1980* by Erik Høg, 28 May 2008

[34] Bulletin of the American Astronomical Society, Vol. 25, p1498 (1993)

[35] Matra Marconi Space is now known as EADS Astrium, but its name at the time is retained throughout this account

[36] New Scientist, Issue 1678, 19 August 1989

[37] Commentary on the failure was given, by reporter Peter B. de Selding, in the 11 June 1990 issue of Space News *Europeans Tussle over Motor Failure*, and in the 16 July 1990 issue *Hi-Shear Part Blamed for Hipparcos Failure*

[38] Each card, of size 7-3/8 × 3-1/4 inches, encoded up to 80 characters over its 80 columns, each represented by rectangular holes in each of 12 punch locations

Notes

[39] 59th Meeting of ESA's Science Programme Committee, 26–27 February 1991

[40] Quoted in *The Observer*, 24 March 1991

[41] Sir George Darwin, astronomer and mathematician, was a son of Charles Darwin

[42] Andrew Derrington, 'The Ruler Strikes Back', *Financial Times*, 29 March 1997

[43] The 'Messier' objects are nebulae and star clusters catalogued by French astronomer Charles Messier (1730–1817). Viewed with the scale and resolution of modern telescopes, many are seen to be galaxies like our own, but at enormous distances. The designations M1 to M110 are still in use by astronomers today

[44] Details are given in *A Short Biography of Jan Hendrik Oort* by J. Katgert–Merkelijn (Leiden Observatory, 2000). The obituary quote was by Chandrasekhar

[45] Stars on the main sequence are referred to as dwarfs. Such a label is confusing because it implies that they are smaller than 'normal' stars, while they are normal for stars on the main sequence. Giants are, in contrast, very large compared to the main sequence stars of the same temperature or spectral type. White dwarfs and brown dwarfs *are* much smaller than normal stars on the main sequence

[46] In science, the more fundamental Kelvin scale of temperature is generally used. The difference between them is 273 degrees, and in this context can be ignored

[47] A history of attempts to date the Earth is told in *Ancient Earth, Ancient Skies* by G. Brent Dalrymple (Stanford University Press, 2004)

[48] Asteroid identification is conventionally specified by its discovery sequence number, followed by a name authorised by the International Astronomical Union. The first few were named after figures from Graeco–Roman mythology, but as available names ran out the names of famous people, literary characters, and others were used. More recently, discoverers have often bestowed the names of other astronomers, from which practice I have my own name carved upon the otherwise unremarkable 10969 Perryman, discovered by the prolific asteroid hunters from Leiden University, Ingrid and Kees van Houten, sandwiched alphabetically between 5529 Perry on the one side, and 9637 Perryrose on the other

[49] 'Tropic' comes from the Greek for turning, indicating that the Sun appears to 'turn back' in its path at the solstices. When they were named, the Sun was in the direction of the constellations Cancer and Capricorn; this is no longer true due to the precessional (wobbling) motion of the Earth

[50] The International Latitude Service was renamed the International Polar Motion Service in 1962, in turn replaced by the International Earth Rotation Service in 1988

[51] The cold period coincided with the reign of Louis XIV (1643–1715). It seems somewhat perverse in the current context that he was referred to as the Sun King. Yet just as the planets revolved around the Sun, so the story goes, so too should France and the court revolve around him

[52] wikipedia/SETI, cited on 30 November 2009

[53] Malcolm Longair in 'Facing the Millennium', *Publications of the Astronomical Society of the Pacific*, January 2001, Vol. 113, pp1–5

Stereo Views

THE FOLLOWING TWO PAGES are reconstructions of details from the Hipparcos catalogue allowing regions of sky to be viewed in stereo, with the distance of the stars visible. The areas of the sky covered are eight by six degrees for each. In both cases, the stars shown are those observed by the satellite. Sizes are shown according to the star brightnesses, and the colours reflect their temperatures, with white corresponding to hotter stars, and red to cooler. The left and right pairs are reconstructed from the satellite measurements so that the resulting stereo effect reflects the true distance of the stars, showing how the stars are really distributed in space.

The first pair shows the region of the Hyades star cluster, at a distance of about 40 parsecs or about 130 light-years. The brightest star, just to the above left of centre, is the foreground star Aldebaran; it lies at about 20 parsecs distance, and is not a member of the Hyades cluster. The cluster has an age of about 625 million years.

The second pair shows the region of the Pleiades star cluster, at a distance of about 125 parsecs or about 400 light-years. The cluster has an age of about 100 million years.

To view the stereo pairs, concentrate on the images from a distance of around arm's length, although a little closer or further might work better. The idea is to focus the eyes on the page, but to 'cross' the eyes so that the right eye looks at the left image, and vice versa. Give some time for the two images to merge into one, and bear in mind that the bright stars in both fields are in the foreground, such that they will be seen 'hanging' in front of the paper. Once you have figured out the effect, which may take some minutes the first time, it can usually be repeated again rather easily.

Jos de Bruijne and Michael Perryman

Stereo image pair of the Hyades star cluster

Jos de Bruijne and Michael Perryman

Stereo image pair of the Pleiades star cluster

Acknowledgments

Hundreds of people played their part in making Hipparcos a success. Many engineers and managers in ESA and industry spent a decade or more engaged in their own crucial contributions. The scientific elements demanded even lengthier commitment down the years—numerous astronomers in Europe dedicated the best part of two decades to their parts in the complex story, the Hipparcos project occupying by far the largest part of their professional lives.

This popular account is not the place to list these detailed contributions, and I can only ask for their understanding in not mentioning most by name; more than three hundred essential scientific and technical contributions are credited in the published catalogue. But as heads of their respective teams, I acknowledge the leadership of project managers Franco Emiliani and Hamid Hassan in ESA, Dietmar Heger at ESOC and, in industry, Michel Bouffard and Bruno Strim who managed the leading industrial teams of Matra Marconi Space and Alenia Spazio respectively.

Astronomers Catherine Turon, from the Paris Observatory in Meudon, Lennart Lindegren from Lund University in Sweden, Erik Høg from Copenhagen University, and Jean Kovalevsky from Grasse in France, played major parts in the early studies, led the four scientific consortia during the 1980s and 1990s, and gained both international recognition for their achievements, and my gratitude for their stimulating collaboration, support, and advice.

The other members of the Hipparcos Science Team for many years—Ulrich Bastian, Pier Luigi Bernacca, Michel Crézé, Francesco Donati, Michel Grenon, Michael Grewing, Floor van Leeuwen, Hans van der Marel, François Mignard, Andrew Murray, Rudolf Le Poole and Hans Schrijver—all played key parts in steering the project to success, as did the many other scientists across Europe within their teams.

Acknowledgments

In preparing this account, my text on the early history of astrometry draws on the cited works of David Goodman & Colin Russell, Michael Hoskin, and Allan Chapman. The decision-making process for Hipparcos through to 1980 pre-dates my involvement in the project, and is largely based on the authorised history of the European Space Agency by J. Krige, A. Russo & L. Sebesta, supplemented by the more recent perspectives of those referenced. For sharing their specific recollections to add to this account, I am grateful to ESA project team members Kai Clausen, Franco Emiliani, and Oscar Pace, and to Dietmar Heger at ESOC.

I also thank all who kindly responded to my requests for the pictures that illustrate the text. Every effort has been made to obtain appropriate permissions in the case of copyright material. Sources and credits are given alongside the images. Portrait photographs are either from my own collection and used with due permissions, or provided by the individuals. Other unattributed figures were prepared for this work by the author.

Various people kindly read parts of earlier drafts, and have been generous with their time and advice. Specifically I would like to thank Adriaan Blaauw, Giacomo Cavallo, Graham Dolan, Ute Eberle, Vincent Higgs, Jean-Claude Pecker, Daniel Reydellet, William van Altena, and Arnold Wolfendale.

I am grateful to Erik Høg, Jean Kovalevsky, Lennart Lindegren, and Catherine Turon for their detailed comments on the full text. François Mignard shared his considerable knowledge of the history of science by providing numerous detailed suggestions on the historical chapters. Roger–Maurice Bonnet generously looked over the entire manuscript, providing some valuable commentary and also some much-appreciated encouragement in pursuing this chronicle through to completion.

I owe a debt of gratitude to my first head of division in ESA, Brian Fitton, for entrusting me with the scientific leadership of Hipparcos at the relatively tender and unproven age of twenty six, while Brian Taylor in ESA provided me with valued support over many years thereafter.

Where I have failed to properly reflect the events or contributions of these and other participants in the making of history's greatest star map, I offer my sincerest apologies, and trust that for all there is some satisfaction in seeing the story related in print, however inadequately. In this context it is a pleasure to thank my editor at Springer, Ramon Khanna, for his assistance in turning it into reality, and his numerous comments along the way.

I leave my most important debt of gratitude until last. Over many years my wife Julia has been a staunch supporter of Hipparcos, or at least my involvement in it. I thank her for her comments and help on this book also, but more profusely for her tireless inspiration and encouragement over the best part of three decades.

Index

Printing and Binding: Stürtz GmbH, Würzburg